Conservation Legacies of the Florence Flood of 1966

Conservation Legacies of the Florence Flood of 1966

Proceedings of the Symposium Commemorating the 40th Anniversary

Edited by Helen Spande

Archetype Publications

First published 2009 by Archetype Publications Ltd.

Archetype Publications Ltd.
6 Fitzroy Square
London W1T 5HJ
www.archetype.co.uk

Tel: 44(207) 380 0800
Fax: 44(207) 380 0500

ISBN: 978-1-904982-44-9

British Library Cataloguing in Publication Data
A catalogue record for this book is available from the British Library.

Typeset by Caz Jones Design, Norwich
Printed and bound in [copy to follow]

Contents

Preface

Ellyn Toscano

I would like to remember two people who symbolize the potential of international collaboration for me. One of these, Maria Fossi Todorow, is remembered not only for her pioneering museum work at the Uffizi, Palazzo Davanzati, and Villa La Pietra, but also as a treasured teacher who brought Italian art history alive to many New York University students. It was Maria who first discussed the idea of organizing a conference commemorating the international conservation efforts with me and we are pleased she was able to see the enthusiastic response and success of the event.

The second is Senator Ted Kennedy, who raised awareness worldwide of the severity of the damage caused by the flood and inspired so many with his visit to Florence in November of 1966. His participation in this conference, 40 years later, demonstrated his continuing commitment to the cause that brought us together, then as now.

The second day of the conference was very generously hosted by Mayor Leonardo Domenici in the Sala dei Cinquecento of Palazzo Vecchio and the grandeur and history of the room gave the closing of the symposium an extraordinary commemorative tone that all present will surely not forget. The concurrent oral history project also involved not just the participating interviewees, but a host of colleagues interested in the history of conservation, including many students from the University of Delaware/Winterthur Program in Conservation. The collaboration with the Harvard University Center for Italian Renaissance Studies at I Tatti, which shared materials from the CRIA Archives, further enhanced the scope of the event.

We were proud to host this important conference at Villa La Pietra, the site of New York University in Florence. Villa La Pietra was bequeathed to New York University by Sir Harold Acton because he hoped it would send its students, scholars, poets, painters, writers, intellectuals, scientists and artists of all sorts to join their Italian and European counterparts in conferences and meetings on a wide range of topics. Villa La Pietra was the ideal place for this conference since the conservation work at the Villa continues the important legacy of international cooperation in conservation which, in many ways, began during the flood.

Foreword

Rebecca Rushfield

I was 12 years old when the Arno broke its banks in 1966 flooding the city of Florence and what I heard from news reports about the rescue effort made it seem – mud aside – like a grand and meaningful adventure. And one that illustrated a belief shared by many young people in the 1960s that if they worked together they could change the world for the better.

Youthful idealism not withstanding, 1966 was not a good year for international cooperation. The Cold War was at its height, the war in Vietnam was escalating and there were ongoing wars in the Sudan, Guatemala, Laos, North Yemen, and between Indonesia and Malaysia. There were coups and civil wars in many African nations and Argentina. There were Irish Republican Army bombings. Martial law was declared in Cuba. And the Cultural Revolution began in China. In stark contrast to this violence, the Florence flood rescue effort was an example of what international cooperation could and should be as both professionals and untrained volunteers arrived from all parts of the world including the Soviet Union to help save the cultural heritage of Florence.

In the decades that followed 1966, the artworks, artifacts, architecture and books damaged in the flood were conserved. The field of conservation changed in great part due to the approaches, techniques and materials developed in response to the flood damages. The Vietnam War ended and the Soviet Union collapsed. And young people became more cynical about how easy it was to change the world.

In July 2003, I interviewed Marco Grassi, a paintings conservator, for the Foundation of the American Institute for Conservation Oral History of Conservation Archives. During the interview, he spoke of his experiences in Florence before, during and after the flood. He told me of a meeting at the Uffizi led by Umberto Baldini on the second or third day after the Arno flooded at which all of the paintings conservators who were in the city – both Florentine and foreign – came together to discuss what should be done with the panel paintings that had been submerged in water and were in dire condition. I was struck by his description of the sense of community that existed among conservators of all backgrounds, ages and outlooks. Realizing that the 40th anniversary of the flood was only just over three years away and knowing that many of the leaders of the rescue effort had died while others were getting on in years, it seemed crucial to record as many of their stories about the rescue effort as soon as possible, both to recapture that special event and to come to understand how that effort led to such great changes in the field of conservation.

In the summer of 2005, following two years of research on the flood and the rescue effort, I contacted Michele Marincola, Sherman Fairchild Chairman of the New York University (NYU) Conservation Center with a very rough idea for a conference on the subject. Fortuitously, at the same time that I was conducting my research and developing the idea for a conference, Michele and Ellyn Toscano, Director of the Villa la Pietra and NYU Global Initiatives were working on plans to hold conferences focusing on conservation-related topics of international significance at the Villa. We combined our efforts and work began in earnest on the choice of speakers and the structure of the sessions.

There were so many aspects of the flood and its aftermath that could be evaluated including new approaches to the treatment of large numbers of damaged objects, the organization of conservation laboratories, the utilization of new materials in conservation, the building of a network of support for a large-scale international effort, the use of untrained volunteers and students in conservation, and disaster preparedness. We knew that we could not cover all of these topics in the two days that the symposium would last so we decided to focus on the international effort, knowing that the Florentines were better able to analyze the local response to the flood. Five sessions were organized: the experience of book and paper conservators; innovations in conservation materials and techniques for frescoes, easel paintings and sculptures; the case of the Italian Art and Archives Rescue Fund; working with students in disaster recovery; and preparation for disasters.

In order to develop a fuller picture of the rescue effort, we felt it was essential to capture the personal stories of our speakers and of other people who had been involved in the rescue effort – where they lived, how they got around the city, and what they did when they were not working. We thus developed a program of oral history interviews that were conducted by conservation graduate students and had the tapes of these interviews transcribed.

We believe that the symposium papers and oral history excerpts in this volume will prove to be a valuable resource for anyone who wishes to understand both the field of conservation as it was in 1966 and how it was revolutionized by the Florence flood.

Introduction

Michele Marincola

Ask any American art conservator of a certain age why she or he originally chose this field to pursue and the chances are the answer almost invariably will be: 'because of the 1966 flood in Florence'. Although there is still an acute awareness of the devastating effects of the flood, the lessons learned from the experience have been the least explored. The catastrophe revolutionized the field of art conservation as no other single event had done, and served as a catalyst for the development of new materials and methods of restoration, particularly for mass and large-scale treatment. The valiant efforts of the Italian people – in particular the Florentines themselves – in saving the artistic patrimony of this great city have been documented and celebrated elsewhere. This book presents papers from some of the conservators, restorers, art historians, former mud angels and fundraisers who participated in the recovery efforts, assessing their experiences in the light of the past 40 years. Most of the contributors were first responders to the crisis or arrived in Florence in the weeks and months after the event. They represent the wide range of nationalities, abilities, philosophies of conservation and training that was seen in the field of conservation at the time – and indeed is still evident.

The papers discuss treatment methods that were devised or perfected in response to a mind-boggling scale of damage: more than 300 panel paintings, 800 canvas paintings, acres of wall painting, furniture, sculpture, hundreds of thousands of rare books and vast tracts of archives were wet, filthy and in need of immediate treatment. Also discussed are experiences –some successful, some less so – with the use of synthetic resins as consolidants and varnishes; the refinement of the strappo technique; the development of methods of salt removal from frescoes and oils and stains from stone; and perhaps most significantly, the development of new techniques for the mass treatment of works on paper and books. Marilyn Weidner, a paper conservator from the United States, filed a report on the work done on books, archives and individual works of art on paper at the Biblioteca Nazionale Centrale in June 1967 that is prescient in its awareness of the historical and professional significance of the efforts:

> Because of the international character of the situation in Florence, results of tests conducted here could very well prove of tremendous importance not only to the preservation of the flood-damaged materials but also to books, archives, and works of art on paper throughout the world. The meetings at the Library and the Archives pointed up drastically the diverse opinions in regard to quality and permanence in the materials we use and the lack of thorough investigations of these materials throughout the world.

Of course, the lessons learned extend far beyond a list of approved materials and techniques, or of approaches to avoid. We want to know how to be better prepared to face such a disaster again, and (better yet) how to mitigate the risk of damage occurring in the first place. As Prague, Dresden, New Orleans and Baghdad have demonstrated, cultural property has continued to be imperiled by natural and man-made cataclysmic events.

I would like to acknowledge the debt we owe Rebecca Rushfield, who conceived the idea of the symposium and who worked tirelessly to realize it in its form. We are grateful to the foundations and individuals who through their generous support made the symposium possible, including the Samuel H. Kress Foundation, the Friends of Florence, the Booth Heritage Foundation, and the Andrew W. Mellon Foundation, which supported the attendance of graduate students from the American conservation programs. Most of all, I wish to thank New York University and the amazing staff at the Villa La Pietra, Florence for supporting the idea from its inception and making it happen.

Acknowledgements

New York University's symposium on the Florence flood held in November of 2006 was only possible through support and sustained interest from both sides of the Atlantic Ocean. Financial support for the conference and the publication of the proceedings came from the Mellon Foundation, the Friends of Florence, the Samuel H. Kress Foundation, and the Booth Family Foundation. Marco Grassi and Malcolm and Carolyn Wiener also contributed essential support that made the event possible.

From the initial planning stages of the conference and through the course of defining the list of speakers, many shared their memories with us and contributed to the shaping of the event. Valuable advice from members of the Florentine community as well as staff expertise from many departments of New York University saw the conference through to its realization. Our appreciation goes to all of these individuals who supported and encouraged this endeavour all the way through to the publication of these proceedings. Our thanks go to Cristina Acidini, Lisa Ackerman, Carla Bonanni, Simonetta Brandolini d'Adda, Jamieson Bunn, Costanza Capaccioli, Gail Cimino, Joseph Connors, Roberto D'Alimonte, Ilaria della Monica, Cesar Francia, Marco Grassi, Simone Grant, Scott Hughes, Shirin Khaki, Myron Laskin, Alta Macadam, Michael Rocke, Bruno Santi, Andrea Scavetta, Alessandro Sidoti, Joyce Hill Stoner, Malcolm and Carolyn Wiener, and Richard Young.

FOR IMMEDIATE RELEASE
November 11, 2006

CONTACT: Laura Capps/ Melissa Wagoner
(202) 224-2633

Senator Edward M. Kennedy
Remarks at Symposium
On the 40th Anniversary of the Arno Flood

Thank you, Ellyn Toscano for that generous introduction, and I commend you for the excellent job you're doing for NYU in Florence as Director of Villa La Pietra and Global Initiatives. It's a remarkable opportunity for NYU students to broaden their educational experience, and for many of them, it's the beginning of a lifelong affection for the people and culture of Italy.

It's also a privilege for my wife Vicki and I to be here with Mayor Domenici. He has the enviable responsibility of presiding over one of Italy's, and the world's, crown jewels – the city of Florence, and I admire his impressive leadership. Each time I visit this city, I'm greeted warmly and treated with the incomparable Tuscan hospitality. Thank you all so much.

I was privileged to visit Florence a decade ago on the occasion of the 30th anniversary of the flood. It was my first visit to the city after the flood and much had changed. Having seen some of the photographs of my earlier visit, I realize that I had changed a lot too! It is an honor to be with you again.

No one can visit this magnificent city and not be humbled by the work of the great leaders in history and culture who walked these streets, whose talent embellishes the churches and villas, whose paintings and sculptures are some of the best known and most revered in all of human history.

It's a special honor to be here today with the Mayors of Dresden, Prague and New Orleans and experts from around the globe who have come together to share ideas and practices for dealing with the disasters caused by immense floods and how to best protect human life, natural resources and cultural treasures when massive flooding occurs.

I commend you all for the vision and hard work that created this Symposium. On this fortieth anniversary of the Flood, it's an impressive expression of international unity, and I commend each of you for your commitment. The city of Michelangelo and da Vinci – this magnificent city which gave birth to the Renaissance – has brought us together once again to inspire new ideas and discuss our vision for the future and the best means to achieve it.

Florence has been called the Athens of Italy. I've just arrived from Boston, which is sometimes referred to as the Athens of America. There are many parallels between these two great founding cities. Neither is a national seat of government, but both are highly esteemed as national centers of culture and learning. Their influence far transcends the size of their population – because they remain deeply committed to the highest levels of human understanding and achievement.

One of the most impressive partnerships between our two cities is I Tatti. As many of you know, Bernard Berenson moved to Florence after graduating from Harvard University in the early 1900's. He came with the financial support of Boston's Isabella Stewart Gardner, who strongly supported his passion for Italian painting. With his brilliant mind and years of study, he rose to prominence and became one of the most renowned scholars of Italian art history.

When he died, he bequeathed his villa to Harvard to carry on his important work. Villa I Tatti remains today a cherished and respected extension of Harvard's main campus, a vital part of Florentine culture and scholarship, and a cherished link between our two magnificent cities.

I'm proud as well to note that the City of Boston is doing its part to help restore the extraordinary artistic heritage of Italy. A month ago, our city's Museum of Fine Arts formally returned 13 archeological treasures it had acquired in good faith, but that had turned out to have been illegally excavated here in Italy and sold through unscrupulous dealers. The head of the museum signed an agreement pledging full cooperation in halting the plunder of such priceless works of art in the future.

I am also proud of the support of the American people for the people of Florence at the time of the Flood.

When the Arno River raged over its banks forty years ago, it damaged or destroyed countless buildings, businesses, and schools throughout Florence. It changed the lives of all the people in this region, but it never extinguished their spirit.

I'll never forget my own trip to Florence at that time. I later wrote about it and this is what I said, "*I remember that I was in Geneva at a conference on refugees and I wanted to see what had occurred, so I flew in to Florence for the day. I got to the library about 5 PM and I looked down into the flooded area. There was no electricity and massive candles had been set up to provide the necessary light to rescue the books.*

It was terribly cold and yet I saw students up to their waists in water. They had formed a line to pass along the books so that they could be retrieved from the water and then handed on to a safer area to have preservatives put on them. Everywhere I looked in the great main reading room, there were hundreds and hundreds of young people who had all gathered to help.

It was if they knew that this flooding of the library was putting their soul at risk. I found it incredibly inspiring to see this younger generation all united in this vital effort. It reminded me of the young people in the United States who responded with the same determination as they became involved in the civil rights movement.

I was still shivering as I boarded the plane that took me back to Geneva, but I couldn't stop thinking of the impressive solemnity of that scene – of all those students, oblivious to the biting cold and the muddy water, quietly concentrating on saving books in the flickering candlelight. I will never forget it." And indeed I never have. How inspiring to see the countless volunteers from around the world who had come here to help in the recovery. Each of them understood the importance of the city's cultural resources. They knew what was at stake, and, despite the tragic and devastating conditions, they worked brilliantly to save the irreplaceable books, paintings, and cultural treasures of the city. They were called Mud Angels, and I am proud to be an *Angelo del Fango* as well.

The Mud Angels earned a page in history and a special place in the hearts of the people of the world. Their extraordinary effort will forever be remembered as a brilliant example of caring and triumph in the face of daunting adversity.

The memory of that tragedy brings us together today, but it is a remarkable victory that we celebrate. Many of the great books, manuscripts, and works of art that might have been lost were saved.

My sister-in-law, President Kennedy's wife Jackie, led the American effort as President of the Committee to Rescue Italian Art, and her nationwide campaign raised millions of dollars to provide essential salvage and restoration efforts. It was her first public involvement after the loss of my brother. She understood that the great works of art cherished in this city are important to the world as part of the great heritage of our common humanity.

As President Kennedy once said, quoting the great American poet Robert Frost, "the nation which disdains the mission of art invites ... the fate of having 'nothing to look backward to with pride, and nothing to look forward to with hope'." Florence has much to look forward to with hope.

The healing and restorative process that began in the muddy wake of the flood has ensured that Florence today is once again a resplendent city – both artistically and economically.

That history gives me great hope for the future of our beloved American city so recently ravaged by floodwaters. Sadly, the tragedy that struck New Orleans was both inevitable and avoidable. Years of environmental neglect had led to erosion of the natural wetlands protecting the city. Engineering miscalculations had created levees that were inadequate to hold back the rising waters. Warning signs had developed and intensified over many decades, but no preventive steps were taken.

Our unique city paid an enormous price for those years of neglect and reckless oversight. To make matters worse, even after the disaster struck, the response of our national government was shockingly inadequate.

Although there is still much work to be done, New Orleans has begun to heal and rebuild, and I believe it will once again be the city of culture, jazz, and joyfulness we all knew and loved.

My wife Vicki is a native of Louisiana and was educated in New Orleans, her second home. As part of the Katrina recovery effort, she led a campaign to raise funds to save the Louisiana Philharmonic Orchestra. With the help of Michael Kaiser of the Kennedy Center for the Performing Arts in Washington, they raised enough to enable the orchestra to resume its performances next season.

There are countless other stories of how our culturally vibrant city is once again finding its strength in its cultural roots, and is beginning its long journey back to prosperity. It's an enormous challenge, but the success of Florence – and Dresden and Prague – inspires us.

The leadership of the Medicis, the genius of Botticelli and Caravaggio, Puccini and Verdi, and so many others inspires us today. Our shared love of their great works also inspires us to seek a new renaissance of ideas and achievements – not to advance a party or political cause, or to enrich the already rich, but to ensure that we leave this world a better, fairer, and more beautiful place than we found it.

Americans cherish their close relationship with Italy, not least because so many millions of Americans are of Italian ancestry. Sons and daughters of Italy have contributed in countless ways to our democracy and enriched our history and culture.

I'm reminded of that heritage every day as I walk through the Brumidi corridors of the U.S. Capitol building. The brilliant murals capture moments from America's early struggles for independence, and they've earned a place for Brumidi among America's most beloved artists.

Today, our shared legacy binds us more tightly than ever. It's a friendship and partnership that mean a great deal to me personally.

As we meet to mark the 40th anniversary of the Flood – an important time in Italy's history, we also meet at an important moment in America's history. American voters have rejected those who sow divisiveness and doubt, and embraced anew a positive vision of America in partnership with her historic friends and allies.

The people of the United States understand that we and our European friends share a common spirit and common values, and that together we can make the world safe for progress.

America will always be a strong partner with Italy. Across the years, we have almost always been united in our hearts, our heritage, and our hopes.

I know there is much more that we can accomplish together. I look forward very much to being a part of the Italian-American partnership in the years ahead, and to enjoying the unique friendship between nations who believe in achievements that embrace wisdom, respect, compassion and beauty.

I congratulate you on undertaking this important symposium. I know that the work you accomplish here will have great significance in the years to come. Well done.

In closing, let me say to Mayor Domenici and Ellyn Toscano who graciously invited me to come here, *'mille grazie'*; and to the former Mud Angels and other participants in this important conference, I say, '*Bravissimi!*'

Oral history of the Florence flood: personal recollections

Introduction

While the formal papers in this volume can provide an understanding of the conservation work that took place in the aftermath of the Florence flood, they do not provide a full picture of what it was like to be a participant in the rescue effort. For that, we must turn to their personal stories, many of which were recorded in oral history interviews conducted by students of the Winterthur/University of Delaware graduate program in art conservation on 10–12 November 2006.

- *The Florence Flood is one of those events for which each person vividly remembers where they were and what they were doing when they heard the news:*

Marco Grassi: It was raining all over Europe: heavier and more than normal. The papers had been full of features about the swelling rivers, so it was clear that this was going to be a long spell of bad weather. At the gallery [in Lugano, Switzerland], an employee who acted as a custodian came in to close up and told me he had heard something on the radio about Tuscany being particularly hard hit by overflowing rivers. The following morning when I arrived at the gallery he informed me that he had just heard a radio bulletin stating that in Florence the water had reached the first floor of Palazzo Pitti. Of course, I immediately turned on the radio – the Arno had already burst its banks and was, in fact, flooding the city. This was a real emergency. I immediately set off in my car but it took me about 15 hours to get to Florence. The major access roads to the city were blocked by the military and emergency equipment that was being sent in. *(In 1966, Marco Grassi was a conservator in private practice whose time was divided between Florence and Switzerland.)*

Andrea Rothe: It had been raining constantly. The night before the flood, my wife and I were driving across Ponte Santa Trinita when suddenly, out of the blue, she said 'Does the

Arno ever overflow?' And I answered 'No.' Little did I know then. It rained intensely the whole night and in the morning there was no electric power. Although it was raining, I had decided to go to work. Because 4 November was a holiday – *Giorno delle forze armate* [the day of the armed forces] when the Carabinieri have their big parade – there would be no one around, and I could work in peace. Early in the morning, as I drove down Via Bolognese past the Villa Acton, I noticed that many of the retaining walls along the street had collapsed. When I reached Piazza San Gallo in the city, traffic was going every which way. Although I am used to the chaotic Italian traffic, I sensed that something unusual was going on. It was very eerie. When I reached the Arno, I saw to my horror that the raging waters had reached the retaining walls and were starting to spill over. I hadn't heard the news and I didn't realise until that moment that the Arno was actually overflowing. The water was starting to spill over the steps at the end of the Piazzale degli Uffizi near the Arno. *(In 1966, Andrea Rothe was living and working in Florence.)*

Joe Nkrumah: One beautiful evening, while having dinner with friends in Valencia, eastern Spain, we heard on the TV that the Arno had flooded its banks. Knowing what Florence was like, I could imagine the extent of the damage. I tried hard to make arrangements to go to Florence to help but communications had been knocked out. *(In 1966, Joe Nkrumah was working at the Pitt Rivers Museum in Oxford, England.)*

Peter Matthiesen: The news came through on 5 November. I think the *Daily Telegraph* actually carried the best report of the flooding of Florence. There was a big buzz at the Courtauld Institute that day. We lobbied our professors by asking 'Don't you think we should be doing something about this?' Their reaction was that the problem had probably been exaggerated and not very serious. That evening I telephoned some friends in Italy and asked them what the state of play really was – I was aware that much of the city is low lying below the Arno. My friends provided fairly graphic reports about what was being shown on Italian TV. *(In 1966, Peter Matthiesen was a student at the Courtauld Institute in London.)*

Natalie Brooke: We heard about it on the television, on the radio and in the newspapers. Sir Ashley Clark (who had for nine years been the English ambassador in Rome) and the Earl of Harewood (the Queen's first cousin) – both connected with the British Italian Society – wrote an appeal letter that was published in *The Times* alerting us all to the problem. *(In 1966, Natalie Brooke's husband was Secretary of the Royal Academy in London.)*

- *The Florentines took immediate action*:

Marco Grassi: People just turned up. It was amazing. The first day I was there, museum conservators came to help as well as a number of private conservators, of which I was one.

Dorothy Cumpstey: Professor Casamassima (Director of the Biblioteca Nazionale Centrale di Firenze) had been a leader in the Resistance so he had that kind of presence of authority.

He was having difficulty getting across the river to reach the library. When he finally did get across, he had to decide whether or not to open the doors to let in the students who were waiting. They pleaded with him: 'Please sir. Can't we help?' So he opened the door and let them in.

- *When people outside of Florence heard of the flood and the damage to its art and architecture, some left immediately to help with the rescue effort. They found devastation:*

Peter Matthiesen: I rang my brother-in-law whose family owned a large, extensive gravel pit company so they had a lot of heavy machinery. I explained what had happened and that I wanted to go to Florence. I asked if he would like to come with us as I thought it would be very exciting. He persuaded his father to let him have a Land Rover and a big industrial pump. I recruited a bunch of students and we travelled to Florence non-stop. We argued our way through the cordon. Fortunately we had a four-wheel drive vehicle – only a vehicle of this type could possibly have got through. The mud was indescribable. The cold was appalling – it went right through our bones. The damp and the stink is something that lives with me to this day – the smell of dead cats, rotting flesh, decomposing material and sewage all over the place. Fuel oil was spread all around the city. We drew up in front of the Uffizi and I asked 'I am here – what can I do?' They looked at me with horror because it was by now either the 9 or 10 November – less than a week after the flood. I looked around the basement of the Uffizi with all the blackened boxes of stuff in it … the place was absolutely awash with water. I queried why it had not been pumped out and was told that they hadn't got any pumps. I said 'I've got a pump!' So the first thing we did was to pump out the Uffizi's basement.

- *Other foreigners came later as part of an organized effort*:

Lucilla Kingsbury: In November 1966, I went to Florence, as did several of my friends, because we had spent time in city. Initially I stayed just for a fortnight and did little bits of this and that. Then I was asked if I could come back as I would have an important role to play. With permission from the Courtauld, I returned to Florence and stayed for a year. *(In 1966, Lucilla Kingsbury was a student at the Courtauld Institute in London.)*

Mette Bjarnhof: We wanted to do something as we were all trained restorers. It was not just a matter of removing things from the mud. We [Scandinavian conservators at the Nordic Centre established in Florence] started in 1967 and stayed for three years. *(In 1966, Mette Bjarnhof was a conservator in private practice in Denmark.)*

- *People came from all parts of the world:*

Joe Nkrumah: People came from Hungary, from Poland, from wherever – I gave them their assignments and oversaw what they were doing.

Karel Stretti: Czech restorers worked in Florence from 1966 to 1970/71 as a state initiative organised by the Ministry of Culture. The ministry sent out professional restorers, paper and graphic art restorers and students of painting restoration with their leader Professor Slansky: about 16 [painting] restorers and students of restoration, and about 10 restorers of paper.

Dorota Niesiolowska: Conservators from Poland worked in Florence for a total of two years in two groups. The first group (1967–1968) consisted of Ewa Wolska, Jerzy Wolski, Gabriela Lipsko and Krystyna Dabrowska. The second group (1968–1969) comprised Wojciech Kurpik, Anna Grzesik, Marta Zurawska, Ewa Wolska and me. The initiative was taken by Stanislaw Lorenz, then Director of the National Museum, Warsaw. All were painting conservators who worked at the Fortezza da Basso.

- *Conservators made the acquaintance of their colleagues from other countries*:

Mette Bjarnhof: We had known perhaps one restorer in Sweden, one in Finland and one in Norway but suddenly we were among hundreds of restorers! We could exchange experiences not just with other Scandinavian restorers but also with the Swiss, Americans and Germans with whom we were working.

- *Conservators had their first exposure to materials and techniques favoured in other countries*:

Lucilla Kingsbury: One of the great benefits of this time was the groups of conservators who came from all over the world. When they had arrived as a group, they tended to work as a group and often used the materials that they used at home which had never really been seen in Florence. One particularly exciting innovation was the use of sturgeon glue introduced by the Russian group. It quickly became the preferred material for many people.

- *Politics and international conflicts were put aside*:

Erling Skaug: While we were in Florence in the autumn of 1968, the 'Velvet Revolution' took place in Prague [the 'Prague Spring']. Suddenly Soviet and Iron Curtain forces rolled into Prague. The Czech group and the Polish group [working in the Fortezza da Basso] therefore became official 'enemies'. So they took a day off and got drunk – together. That was really something. *(In 1966, Erling Skaug was an intern at the Istituto Centrale per il Restauro in Rome.)*

- *Despite the camaraderie among conservators, cultural differences created some difficulties*:

Henrik Bjerre: The Italians thought that the foreigners coming into the Florence had created a kind of second flood of people trying to take over, trying to teach them what to do. It was almost the world versus Florence. *(In 1966, Henrik Bjerre was working at the Statens Museum for Kunst in Copenhagen, Denmark.)*

Anne Pelikan: What really impressed me at the time was the difference in the working culture in Italy compared to the United States. The people I was working with (helping out with the flood) were mainly those who, after graduating, had then obtained employment. This particular group was trained for something very specific that they would do for their whole lives. To the Italians, the menders were menders and the washers were washers. Because I was a foreigner I was given special treatment which made me feel a little uncomfortable. I was allowed into all the different areas [of the book conservation laboratory in the National Library]. There were three or four male binders who were trained by Tony Cains; the sewers were women. The two men who carried out the gold tooling had real status – they showed up for work in suits but then they took off their jackets and put on their smocks. There was a real stratification. *(In 1966, Anne Pelikan was working in the Registrar's Department of the Museum of Modern Art in New York.)*

- *However, in the initial salvage effort, everyone did the same work and there was no differentiation among workers:*

Marco Grassi: Everyone would report in the morning – volunteers and some of the employees. Some people had experience but other inexperienced volunteers just turned up. [Giuseppe] Marchini would give them a location and armed with rolls of paper and buckets of Paraloid they were despatched to an oratorio or a sacristy or a convento and so forth. These teams fanned out to do papering and once this had been done, the objects would be taken as quickly as possible to the Pitti Palace where the Limonaia was set up as a gigantic humidity chamber. I was involved in this phase for some time but not for long as these teams were very efficient.

Lucilla Kingsbury: We tended not to carry through work on any particular object – we were asked to do whatever needed to be done that day.

- *Treating as many objects as possible in as short a time as possible was the objective:*

Anne Pelikan: A lot of work was done very speedily but it was set up in a way to ensure safety of the items. It was phenomenal being able to do such a volume of work in a non-destructive way.

- *The great volume of work was possible because systems for mass treatment were put into place:*

Andrea Rothe: In the beginning, there were so many people directing that it became quite chaotic. But fortunately, Umberto Baldini, the State Inspector of Conservation, soon took charge of the situation. The paintings were systematically disinfected with a solution of Nistatine, an antibacterial solution, and then faced with Paraloid and tissue paper.

Anthony Cains: [The books] were sent up to Fort Belvedere where a sort of triage took place using symbols – we had various symbols that we devised mainly to overcome the language problems. The books were sorted as they came in and then shelved. We in turn went through them to see what could be processed first. *(In 1966, Anthony Cains had a bookbinding workshop at St Albans, Hertfordshire, UK.)*

- *None of the conservators had ever been faced with such a large volume of material in such poor condition and creative collaborative approaches were needed:*

Marco Grassi: There was a general meeting of all the people at the Uffizi on the second or third day headed up by Baldini. The idea of this meeting was to discuss the issue of what to do as an emergency measure. Several things became obvious immediately. The canvases posed rather less of a problem but the panels required immediate attention. It was clear that the wood and all of the painting materials had swelled considerably – sometimes unequally where only part of the painting was under water. But no one could stand up and say 'Yes, I've had a Florentine *trecento* triptych under water for two hours. I know what to do.'

- *New materials or novel uses for older ones were required:*

Robert Feller: I was invited by Leonetto Tintori to go to Florence in February/March 1967 to advise him on problems regarding the care and treatment of frescoes. Because the local restorers/conservators had little experience or knowledge of acrylic or polyvinyl acetate resins, I was asked if I would give a series of talks to the local group. With the help of a translator and a blackboard I gave talks to an assembled group of about 20 on, I believe, four mornings, along with a few handouts hastily put together. *(In 1966, Robert Feller was a chemist working on the problems related to museum preservation at the Mellon Institute in Pittsburgh on a Fellowship from the National Gallery of Art.)*

Peter Matthiesen: We had a long discussion [about how to remove deposits of heating oil from sculptures] when the question was posed: 'What happens when, in an Italian restaurant, the waiter throws spaghetti bolognese down a customer leaving an oily mark?' to which I answered: 'The waiter arrives with a tin of boric acid or talcum powder and covers the mess. The talcum powder absorbs the oil and then they brush it off. We need to do that on a big scale.' It just so happened that I had seen a new product from Germany called K-2FR (basically talcum powder in a toothpaste tube containing solvents) that worked on the same principle. It is sprayed onto the stain and left until it dries. As it dries out it shrinks, the talc cracks, the solvent evaporates and draws out the oil into the talcum powder. We, however, would need industrial quantities. We ventured into the non-flooded part of Florence until we found a university bookstore and bought a hand primer of chemical solvents with all their properties – all the petroleum solvents (Gettens and Stout or an Italian equivalent). We then went to the local chemistry wholesaler to the university and bought three or four little phials of a selected range of chemicals and the equivalent of Johnson's baby powder (called Roberto Bototalco in Italy). Back at the Bargello we made up pastes and started making little poultices that we applied to various stains. We tried carbon tetrachloride with xylene, toluene with acetone (nail varnish remover) and a variety of other expedients including Lissapol. We carried on experimenting until we found a concoction that seemed to work. But we needed to do it on a much bigger scale so we got hold of the *Yellow Pages* and rang around the whole of the north of Italy. Knowing that talcum powder is effectively a dolomitic rock ground up, we search the Dolomites area until we found a company that produced industrial talc. We rang them, explained the problem and ordered a ton.

- *Industrial suppliers supported the conservation efforts* :

Joe Nkrumah: We also liaised with industry because, for some of the materials that we needed, the minimum quantity that the industry would sell us would have lasted for 200 years so we had to call the big shots – they sent us generous samples, some of which lasted quite a long while.

Robert Feller: I was shown a large container of Elvacite – normal butyl methacrylate polymer – high on a shelf. Apparently it had not been used thus far ... I learned later that the container had been brought from Paris by a representative of the DuPont Company who had put it in his car and personally driven it down to Florence.

- *Very little of the conservation work would have been possible without the financial support from abroad*:

Henrik Bjerre: The government of the Nordic countries funded the Nordic Centre: Norway, Sweden, Finland, Denmark in addition to private funding. I believe a lot of money was collected in Denmark from the sale of special stamps etc.

Nicolette Bingham: A painting, *Femme couchée lisant*, donated by Picasso to the Committee to Rescue Italian Art was sold at an auction for its benefit during the international satellite programme *Bravo Picasso*. It was presented by the National Broadcasting Company on 5 February 1967. We [the Committee to Rescue Italian Art] received contributions from so many different countries including Micronesia but Italy, of course, was our biggest contributor by a long shot. We had something like $5 from Belgium, $10 from Chile. *(In 1966, Nicolette Bingham was the assistant to Bates Lowy, head of the Committee to Rescue Italian Art.)*

Kirsten Aschengreen Piacenti: All the foreign funds helped to set up the conservation laboratory because, of course, there was no money for that sort of thing. I happened to meet people that I knew from London: Sir Ashley Clarke, who was the Chairman of the Italian Art and Archives Rescue Fund, and Nicolai Rubenstein, who had been my professor of medieval history in London. They were wonderfully generous and sent money to set up laboratories. *(In 1966, Kirsten Aschengreen Piacenti was living in Pisa but cataloguing the ivories in the Museo degli Argenti at the Pitti Palace.)*

Natalie Brooke: We were able to send money direct to Perry Gordon, the British Consul in Florence, or any of our restorers who were working there at that time. I remember one exciting day in the National Gallery when we were told that there were two planes filled with blotting paper which was needed at the Biblioteca Nazionale. I remember authorising them without prior authority from anyone else – I just said 'Go.'

- *Publicity about the destruction and about the conservation work was necessary for fundraising although it could and did get in the way of work:*

Lucilla Kingsbury: You could see people who had just arrived and who knew nothing about the situation having their heads turned by journalists and spending their time giving interviews. It was extremely disruptive having journalists, particularly cameramen, visiting (as they frequently did with all their lights) when we were working on the paintings. Everything had to stop. On the day of the move of damaged paintings from the Limonaia to the Fortezza da Basso we were all issued with white overalls (that no one ever wore normally) and were told to stand by our pictures and hold various bits of equipment because there were some very important visitors due. It was important we should be seen doing these things so we duly obliged. Unfortunately there was a tremendous thunderstorm – thunder and lightning and absolutely horrific rainfall – whereupon it was discovered that the roof of the Limonaia wasn't really quite all it might have been. And water came pouring through. So, all the paintings on tables had to be hastily moved out of the way.

- *While work was the focus of the rescue workers and conservators' lives, they had to deal with the realities of daily life in Florence. For the first responders to the flood, life was difficult*:

Andrea Rothe: There was no light, no water, no phones, and no heating anywhere.

Peter Matthiesen: There was no food – you couldn't buy any food. There were no restaurants – everything in the centre of the city had completely gone.

Kirsten Aschengreen Piacenti: I had the advantage of going home each day to Pisa to a house that hadn't been touched by the flood. In my fully functioning bathroom, I could have a tub full of hot water after a day in Florence.

- *As the rescue effort became more organised, conditions improved for the workers:*

Peter Matthiesen: We [English student workers] went to the fantastic surroundings of the Villa Ombellino, just below the American School up on a hill outside Florence (South). It was an incredible English villa owned by an unbelievable woman, Violet Trefusis, whose mother had been the mistress of George V. We lodged in a disused wing of this colossal villa which had gardens and a view that looked all over Florence, down over the Porta Romana … It was a magic time.

Henrik Bjerre: We had a wonderful time. We were well paid by the government. We had the best seats at the opera. We could travel easily to Tuscany ... in groups of Scandinavians and Austrians, Germans and English.

Peter Matthiesen: In the evenings, we young restorers went to a little trattoria called Otello's (run by a Communist): three courses with a quarter carafe of wine cost 500 lira (50 cents/half a Euro). It's almost unbelievable.

- *Days were full:*

Peter Mallory: Every day, my regular routine in Florence was get up in the morning, have a cup of cocoa at one of the local bars, go to work until midday and then quickly consume a pizza. From 12.30 until 3.00 pm Steve (Herbert – another American student) and I went off to our 'freelance business' working on a beautiful chapel on the Piazza del Duomo. We would go to the materials dump on the back of Steve's Lambretta, get the materials that we needed and work during our lunch hour. Then we'd work again from 3.00 to 6.00 pm at Santa Croce or whatever. At 6, I'd go rowing for an hour on the Arno. *(In 1966, Peter Mallory was a graduate student in art history.)*

- *Through their work and their relationships with colleagues, the rescue workers found their time in Florence to be deeply satisfying and meaningful*:

Peter Mallory: The '60s was a wonderful time because we all believed we could make a difference. And in our own way, we 'mud angels' – I use the term loosely because there was no mud left by the time I arrived before Christmas – really did get a sense that we were making a difference. The success of my work and the work of all my fellow mud angels is that 40 years later there is nothing to show our children. To their eyes, it is as if the flood had never happened.

Editor's note

The verbatim accounts have been edited for clarification purposes.

An account of the flood and the days that followed

Sandro Pintus

It had been raining hard for a number of days but we all knew that 4 November, a public holiday for the annual celebration of the armed forces (*Giorno delle forze armate*), was always accompanied by a downpour. Traditionally it was a wet time of year. The last week of October 1966 had not been so bad – Florence had even hosted an English week – and the autumn had seemed no wetter than usual.

But in the early days of that terrible November, the rain never stopped and the Florentines grew increasingly uneasy. There were rows of umbrellas to be seen up and down the Lungarni and along the bridges as people peered anxiously into the dark and swelling waters. From Piazzale Michelangelo, the scene was not at all reassuring. The long flow of muddy water threatened to burst its banks, and the onlookers, powerless beneath their umbrellas, hoped that somehow tragedy would be averted. Surely the river, which had served the city so well for centuries, would take away that swollen mass of water, the accompanying mud, and accumulated debris?

Alarming news reached the city on the evening of 3 November. In the Mugello, to the north of the province of Florence, and in the Casentino and upper reaches of the Arno valley in the province of Arezzo, all the small rivers had burst their banks. Towards midnight the Arno began to overflow into the Casentino and the Valdarno. Flooding closed the Autostrada del Sole and the railway line between Arezzo and Rome in the area around Incisa in Valdarno. Montevarchi, Figline Valdarno, Incisa, Rignano, Pontassieve, Le Sieci and Compiobbi were all under water.

During the night of 3 November, the mayor of Florence, Piero Bargellini, was woken by a call warning him of the danger to the city. By 2.30 am, the streets of Figline Valdarno, only about 30 kilometres from Florence, were covered by two metres of water. In the surrounding countryside people took refuge on the roofs of their farmhouses while the water moved ever closer to Florence.

But how had this happened? Meteorologists explained it in terms of the abnormal conditions which led to 25% of the annual rainwater falling in 24 hours – more than 190 mm.

In the Florence area the annual average is 823 mm. On those two days in 1966, the whole Arno valley experienced similar dramatic rainfalls. The Osservatorio Ximeniano, which for centuries had recorded weather conditions in Florence, reported that atmospheric pressure fell rapidly more than 20 millibars, with a sudden rise in temperature. There was a belief that a cyclone had hit the country bringing a mass of warm and damp air in its wake.

Above Florence, the hydrometer measured the height of the water at 6.90 metres, and the first muddy surge made its entry into the city. In the Santa Croce area, the drains and manholes were unable to cope with the pressure of the incoming water. Further downstream, beyond the Cascine park, the Mugnone flooded the racecourse and stables. Most of the horses were rescued after three hours of desperate work; only about 70 of the 270 animals were lost.

Rain continued during the night, and the situation became worse by the hour. At 4.00 am the dams in the Valdarno, opened to avoid bursting, added to the swell already enlarged by the water from the higher tributaries. It is thought that this massive swell took between three and eight hours to reach Florence, sufficient time to warn the population but perhaps not long enough to prevent panic and chaos.

The lower parts of the city were soon covered in muddy water littered with cars, uprooted trees, detached doors, tables, chairs, and any other objects caught in the furious course of the flood. The electric clocks in the city stopped at 7.29 am. From then on, the city remained without power; the generator at the hospital of San Giovanni di Dio also failed and an operation to rescue some 200 patients took place by candlelight. Next to the hospital, the army police (the Carabinieri), were grounded as their car engines filled with mud.

By 9.00 am, the areas of Santa Croce and Gavinana were completely submerged: houses, artisans' workshops, cellars and shops filled with mud, and the waters continued to rise. Torrents of water raged through the city at speeds between 40 and 50 km an hour, sweeping away parked cars and battering them violently against buildings, often causing electrical faults and setting off car horns. These horns sounded as the eerie lament of a city suffocating in a mixture of mud and thick, black oil; the baying of desperate dogs added further to the surreal atmosphere. Shortly before 10.00 am, the flood reached Piazza Duomo while waves 2–3 metres high continued to rush into the surrounding streets and squares. Those who lived on the ground floor or in basement flats had, by now, seen all their belongings buried and had sought refuge with neighbours on upper floors or on rooftops. Others, mostly the elderly, failed to seek help because they were unaware of what was happening. While the turbulent slimy lake spread over the city, the river continued to batter away at the Ponte Vecchio

In *4th November 1966: The Flood in Florence* by Silvia Messeri and Sandro Pintus (Ibiskos Editrice Risolo, Empoli, 2006, p. 48), Professor Umberto Baldini, then in charge of restoration at the Gabinetto dei Restauri, gave this description:

> I managed to get into the Uffizi walking through water up to my waist ... in the museum we tried to save any of the works that could be moved. There were paintings by Giotto, Tiepolo, Mantegna, and the *Madonna* by Filippo Lippi. Botticelli's

> *Coronation of the Virgin*, was too big to manoeuvre through the door but we managed to lift it above the waters and out of danger. Having rescued these works, I went to the Vasari Corridor over the Ponte Vecchio with Dottoressa Bechelucci and Dottor Fossi to see what could be rescued there. The whole bridge shook and it appeared to be on the brink of collapsing. We managed to move the pictures to safety before returning to the Uffizi, from where we could look down onto the raging Arno. We could see that the Ponte Vecchio could not take the strain for much longer. Then a lorry swept along in the waters violently bashed into the wall crumbling it like a pack of cards. The opening made by this lorry saved the bridge as the water could pass through the hole unimpeded.

At about 11.00 am, all the Lungarni were completely covered, and the radio in London announced that Florence was about to be lost to the world. American television took up the alarm in their earliest broadcasts about the flood. But while the rest of the world talked about the drama of Florence, there appeared to be a breakdown in communications between the city and the Italian government: neither the President, Giuseppe Saragat, nor the Prime Minister, Aldo Moro, understood the extent of the flood damage and only became fully aware of the situation a few days later.

At 12.30 pm the press agency ANSA communicated the alarm call made by the city mayor, Piero Bargellini. A radio link warned Florentines that the water had reached Piazza Duomo and in some areas was as high as the first floor. The mayor asked the people to bring whatever boats and canoes available to the Palazzo Vecchio. His plea reached the whole world.

The city now had no electricity and no drinking water. The convoy of firemen despatched from Rome were blocked by the flood in the Valdarno. The rain continued, and the water went on rising. The hospitals, too, could no longer function after their generators failed. The Istituto degli Innocenti, built by Brunelleschi in 1429 as an orphanage for abandoned babies which, at the time of the flood housed more than 200 children, including 80 not yet weaned, lost its generator, kitchen and storeroom to the rising water. The Meyer paediatric hospital was also in crisis as the electricity failure closed down the incubators for the newborn babies. Rescued and frightened children continued to arrive.

Alessandra Maggi, who was working at the Meyer at the time, and is now director of the Istituto degli Innocenti, remembers that time:

> Little children arrived at the hospital with their mothers: children who had swallowed the muddy water, or who had been saved from the raging flood, or who had sought safety on the rooftops. They were rescued by boat. They wore expressions of fear and terror. My experience of the flood was through the stories, emotions and eyes of these children and their mothers.

The streets of Florence were more like furious torrents carrying away everything in their path: cars, furniture, trees and people. The metal blinds covering shop fronts offered no resistance, and large doors were flattened. As Sherry Peregrin, an American student, recalls:

> I remember the noise of the water as it rushed under my windowsill. I was trapped in my apartment for two whole days ... I thought of all the people in Florence who lived below the street level, and who could have had no idea of what was happening.

It was still raining, and the water continued to rise. During the afternoon of 4 November, in some places in the centre, the level rose to six metres. When darkness fell, Florence was a ghost city. It was not until 6.00 pm that the waters slowly started to subside, leaving the city to its long, dark nightmare.

When the hydroelectric dams at Levane and La Penna were implicated as possible causes for the deteriorating conditions, ENEL (the National Electricity Board), set about making a detailed report in which they calculated that the quantity of water to have hit Florence was in the region of 250 million cubic metres; 120 million of these came from the higher reaches of the Arno and the rest came from the rivers below the dams. The entire quantity of water was calculated as 400 million cubic metres, and the Arno at its highest was estimated to be carrying between 4000 and 4500 cubic metres a second.

The following day an assessment began of the damage: 70,000 families had been made homeless; almost 6,000 commercial premises destroyed; about 20,000 cars submerged in the mud; 80% of the city's restaurants were wrecked as were two-thirds of Florence's hotels. The city was without gas, water, electricity, telephone lines or a sewage system. It was not only a tragedy for the people of Florence but also for the city's artistic and architectural heritage

Strangely enough, the majority of Florentines who lived outside the city centre failed to realise the true nature of the disaster. The flood was not the result of a great wave, as with a tsunami, but was the result of an 'overflow'. They saw their city slowly filling with water, mud and rubbish, and were unable to do anything about it. Even those who were directly affected could only begin to assess the extent of the damage the following day.

The Florentines soon revealed their natural dignity and courage, rolled up their sleeves and began digging out the mud to rescue whatever they could from the wreckage. And in the midst of all this, something new and quite wonderful happened: groups of young people, in love with the city and her works of art, arrived from all over the world to help the city and its people save what they could. Working side by side with the army and emergency services, who laboured hard for months digging out the mud in terrible conditions, they managed to save thousands of works. This was a real global village in operation, the first real case of globalisation. What is more, Florence became an enormous restoration laboratory combining expertise, methods and techniques from around the world. As Professor Alfio Del Serra, an experienced restorer at the Uffizi, remarked: 'It provided enormous stimulus, mobilising all branches of the discipline, and gave us great strength.'

Invaluable support was provided by a number of show business personalities and political figures who came to Florence: Franco Zeffirelli, Richard Burton and Senator Edward M. Kennedy all visited the Biblioteca Nazionale, which had suffered so badly from the flood. Senator Kennedy described the scene in these words:

> There was no electricity and massive candles had been set up to provide the necessary light to rescue the books. It was terribly cold and yet I saw students up to their waists in water. They had formed a line to pass along the books so that they could be retrieved from the water ... It was as if they knew that this flooding of the library was putting their souls at risk. I couldn't stop thinking of the impressive solemnity of that scene – of all those students, oblivious to the biting cold and the muddy water, quietly concentrating on saving books in the flickering candlelight ... I found it incredibly inspiring to see this younger generation all united in this vital effort (*4th November 1966: The Flood in Florence*, Silvia Messeri and Sandro Pintus, Ibiskos Editrice Risolo, Empoli, 2006, pp. 78–9).

The flood of 1966 was also important as the first to have been recorded with complete photographic documentation, the work of numerous still and ciné photographers. These pictures bring home the extent of the tragedy and preserve the memory of those dark days for future generations. Once the waters had subsided, a count was made of those who had lost their lives: 35 in all, for the most part elderly. An assessment was then made of the damage to the city's artistic heritage: 850 works of art (excluding architecture), were seriously affected – 221 panel paintings, 413 works on canvas, 11 fresco cycles, 39 individual frescoes, 31 detached frescoes, 14 sculptural groups, 122 individual sculptures, 22 sculptures in wood and 23 illuminated manuscripts (*Firenze, i giorni del diluvio*, Franco Nencini, Sansoni, Florence, 1966). The most serious damage was done to the crucifix by Cimabue in Santa Croce; to the frescoes by Paolo Uccello in Santa Maria Novella; to the Botticelli in the church of Ognissanti; to the frescoes in the Spanish Chapel, and to works by Pietro Lorenzetti, Simone Martini and Domenico Veneziano. The Science Museum was totally destroyed, as was the collection of historic musical instruments in the Museo Bardini. Severe damage was done to the weapons collection in the Bargello. The Etruscan section of the Archaeological Museum was almost totally destroyed, and the panels by Andrea Pisano were detached from the Baptistery doors where the doors by Ghiberti also suffered considerable damage.

In the libraries: 1,300,000 items were damaged in the Biblioteca Nazionale. Among the most severe losses were the large volumes of the Medici's Biblioteca Palatina and most of the Magliabechiano collections; valuable manuscripts, sixteenth-century atlases, as well as topographical and other catalogues. In the Gabinetto Vieusseux 250,000 works were submerged, as were 36,000 volumes in the library of the Accademia dei Georgofili. A total of 14,000 volumes dating from the sixteenth, seventeenth and eighteenth centuries suffered in the Synagogue library where hundreds of parchment scrolls of the Old Testament were also affected. Scores of seventeenth- and eighteenth-century *libretti* in the library of the Cherubini Music Conservatory also suffered, as did 9 out of 10 of the university libraries.

To conclude, I should like to quote the words of Senator Edward M. Kennedy during his visit to the Italian Embassy in Washington: 'It would take more than a bit of mud in Florence to bring the statue of David to its knees.'

Editor's note

Translation by Heather Roberts.

The development of mass treatments: an overview of the experience of book and paper conservators

Sheila Waters

Introduction

This paper gives an overview of the setting up of the restoration system at the Biblioteca Nazionale Centrale di Firenze (BNCF) between the end of November 1966 and October 1967, when Anthony Cains took over from my late husband Peter Waters for the following three years as technical director of the system.

I am fortunate to have a great deal of material in Peter's files from which to draw: his published articles, his correspondence and his journal. He was in Florence most of the time during those early months, and I worked with him in the library for a total of three months during 1967. But most of the time I was at home carrying out my own calligraphic design commissions and also acting as a liaison between Peter, the British Italian Art and Archives Rescue Fund, people who were involved in Britain and members of the ever-changing team of binders and restorers.

Peter's journal was in the form of frequent letters to me – over 40 between the end of November 1966 and April 1967 – in which he described the day-to-day happenings. My letters to him kept him informed about everything that was relevant on the home front and boosted his confidence when the problems seemed to be overwhelming. In this paper, I will try to give you a picture of those crucial early months in an abbreviated form. Peter wrote a concise yet very informative article about that period for the *Penrose Annual*, published in 1969. For parts of my narrative I cannot hope to improve on his own words, so I have decided to quote heavily from his article and intersperse with my own words gleaned from his letters. Then, in a way, you may hear his voice through me.

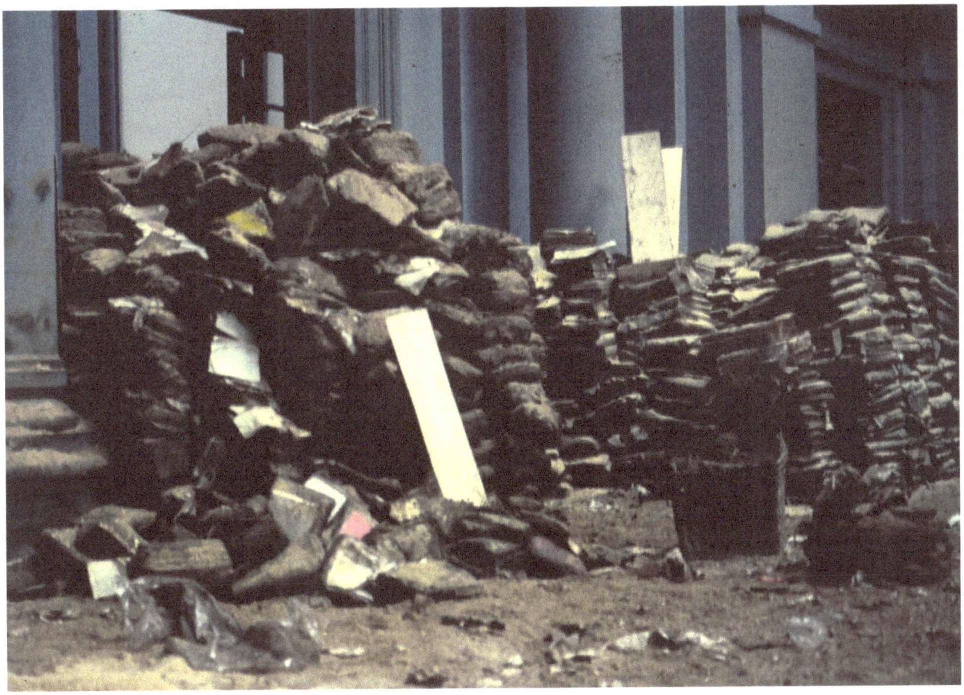

Figure 1 Wet books covered in sawdust.

The call to help and the situation at the BNCF

On 25 November 1966, Peter was phoned by Howard Nixon, then Keeper of Printed Books at the British Museum. He had been contacted by Professor Emanuele Casamassima, director of the BNCF, who needed advice on restoring at least 110,000 sixteenth- and seventeenth-century rare printed volumes, badly damaged by the flood. These included 90,000 of the nucleus of the library, the priceless Palatina and Magliabechi collections.

At the age of 36, Peter was a well-known designer bookbinder and manuscript restorer, in partnership with Roger Powell, 35 years Peter's senior, who had been his mentor while Peter was a student at the Royal College of Art. So Howard Nixon knew that Peter had the knowledge and innovative skills – even though he practiced in a small private workshop – needed to help resolve the vast problems to be faced in Florence. Peter was asked to choose two colleagues and fly to Italy the next day. He chose Anthony Cains, a skilled private restorer, and Dorothy Cumpstey, an expert teacher of binding; fortunately they were able to drop everything to go with him. For the first few nights the two men were housed in monastery cells and Dorothy was in a convent, but later they stayed with the extended team at the Pensione Bretagna, by the Arno, near the library. Without a mandate, they were taken to Forte di Belvedere to look for mould on the books that were arriving there daily having been returned from grain and tobacco drying facilities all over Italy, where vast number of wet books were being dried en masse.

Volunteers were still digging books out of the mud in the basement areas, wiping off the excess with wet sponges and scattering sawdust over them to absorb some of the water. Over 400 truck loads were dispatched to driers. Peter saw, and photographed, high stacks of these wet books both inside and outside the library, awaiting truck transport. In the Fortezza da Basso there were several miles of new shelving and thymol fumigating equipment but no heat and little water. Working conditions were very uncomfortable but the cold discouraged mould growth, so mould was mainly confined to covers and end-leaves.

The state of the dried books

The following quote is from Peter's article in the *Penrose Annual* published in 1969:

> Despite good superficial cleaning before drying and the wise rule by the director at the beginning that none of the books should be opened before arrival at the driers, many of them, arriving at the Fortezza, resembled abstract sculptural forms. Vellum and leather covers were distorted and shrunken. Book stacks in the library have characteristic openings at each end and vellum bindings stored originally at the ends of the stacks had rotted where the water had run through these shapes, combining with the vellum to form a glutinous mass. Other covers were defaced with a combination of mud, sawdust, oil and mould. Sawdust had helped to soak up some water and gelatine, but edges were badly stuck with gelatine, mud and sawdust.
>
> The books had been stood to dry in the heated driers, allowing water and gelatine to drain. The concentration of gelatine was greatest in the openings between the sections, leaving them brittle. Some could be snapped as easily as a freshly-baked biscuit. Covers that had been too hurriedly ripped off had dragged the sewing through the backs of the sections. The water, combined with excessive gluing of the spine when originally bound, had accelerated the damage to the backs of the sections. Some spines and edges suffered greatly through the insensitive use of hard scrubbing brushes. Early 16th century spines suffered least because they had been pasted with starch paste rather than hide glue. The leaves of books bound in limp vellum withstood the flood better than most.

The plan

During those early weeks, Peter used every spare moment to photograph damaged books; there are several hundred in the collection of 1100 slides he took during those months. In an early letter to me, Peter described a high level meeting on 29 November during which it became clear that there was no plan for the future beyond wrapping and storing the dried but very badly distorted books. In that state they would remain unusable by readers. Peter spoke up to question this policy and suggested what must have sounded an outrageous idea: that a colossal pulling, washing, drying and pressing plant be set up to save the majority of the collections in a flat, wrapped state to await mending and rebinding. The effect on the director was dynamic, and from that moment he unswervingly supported the implementation of Peter's plan throughout the following months, no matter the occasional opposition. It was decided then, only four days after Peter and his small team arrived, that the whole series of operations would eventually be carried out in the library itself.

The railway power station had a large number of sinks and a guaranteed supply of hot water; 100 students worked there around the clock in three shifts, supervised by a German family of restorers, separating the leaves of large wet books that had not been sent to the driers. Because the work was so delicate and the inexperienced students were using a technique suitable only for the most skilled hands to attempt, further damage was being done. While Tony Cains supervised students who were dry cleaning at the Fortezza, Peter and Dorothy stayed up for 40 hours at a stretch to monitor the work at the power station. Peter worked on one book himself, finding it took him, with his undoubted skill, a total of eight hours to separate and wash. Very soon after this, he was able to have the work stopped and those particular German restorers went home. The British team was asked to keep that group of students together and train them in the new system of washing, after the sections had been separated while dry – an operation called 'pulling'.

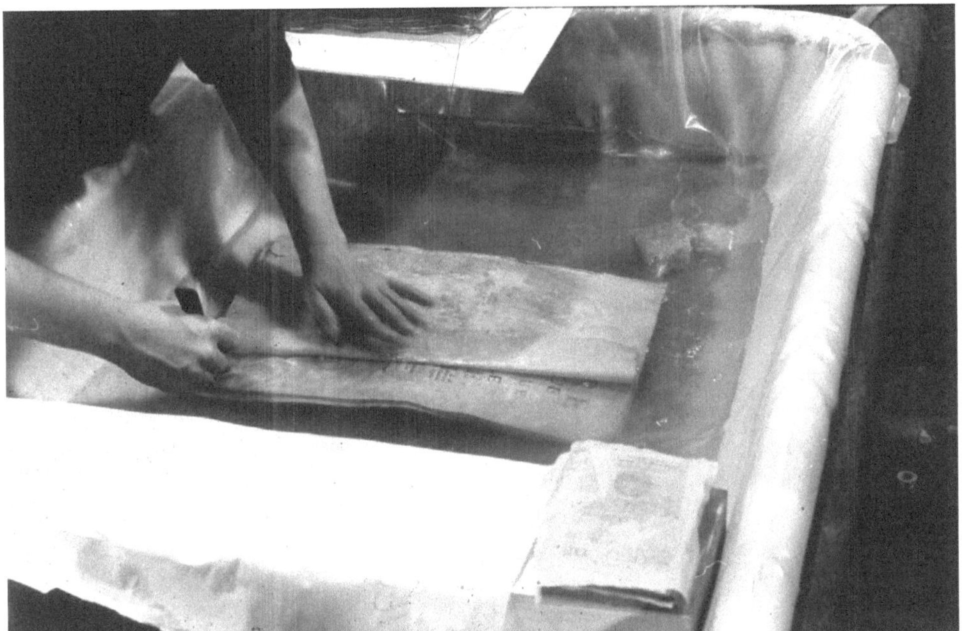

Figure 2 Washing in a power station sink.

Expansion of the team and funding

To continue from Peter's article:

> It became clear that a great deal of additional professional help was needed and we appealed for more British restorers and binders, particularly those with teaching ability. In the following months over forty people worked for periods of from two weeks to several months developing the system and training volunteers. The team included members of the British Museum Stationery Office Bindery and the most distinguished private restorers, binders, and teachers of binding in Britain. This flow of help was organized by Howard Nixon and financed by the British Italian Art and Archives Rescue Fund, whose major effort in Florence became concentrated on the BNCF. From time to time the team was joined by restorers from many other countries. Guidance on chemical problems was given to the team by the British Museum Research Laboratory in collaboration with the Istituto di Patologia del Libro in Rome.

Peter's partner, Roger Powell, arrived early in December, one of the first to come at Peter's special request, accompanied by his wife Rita. Roger was a great help because he could concentrate on the skilled work of selection, deciding on the treatments each book would require. Unfortunately, on 19 January, Peter tripped and fell outside the Fortezza, dislocating the third finger of his right hand. He was concerned that his hand skills would be affected and that he might never be able to play the piano properly again, however, the finger was straightened and strapped in a metal splint – he wrote four pages to me that night in pencil, finishing it with his left hand. He wore the splint for several weeks, never letting it slow him down, and fortunately the finger completely healed.

Preparation for the move to the BNCF

To continue from Peter's article about preparations for the move to the library:

> Meanwhile several rooms in the library itself were being prepared and special equipment designed by members of the team and made by local craftsmen, so that all the work could be transferred and continue under one roof in controlled conditions; permanent workers were gradually to take the place of student labor. An ethylene oxide chamber was installed by the director so that every book would be sterilized before re-entering the library. The move from the Fortezza and the power station took place in March 1967.

Selection for treatments

Peter then continues with precise detail about selection for the treatments necessary for each book on its journey through the system. It is written in the present tense because it continued like that through 1967 and beyond. First he describes the selection process:

> Before a book enters the restoration system, it has to be classified according to its value and the treatment it requires, and details of historical interest are recorded. Symbols were devised during our first week at the Fortezza to make international interpretation easier and designed to warn a student or worker of the state of the book and how it should be handled. For instance, OK indicated that a student could handle a book, whereas the international road 'stop' sign showed that it was particularly valuable and should be reserved for specialist treatment. In the beginning, these signs were written on a slip of paper, but now a detailed printed restoration card and photographs of the original binding and any significant pages, accompany each book with classification and directions for treatment. The card also indicates the history of the original structure by means of an enlarged symbol system with other relevant details.

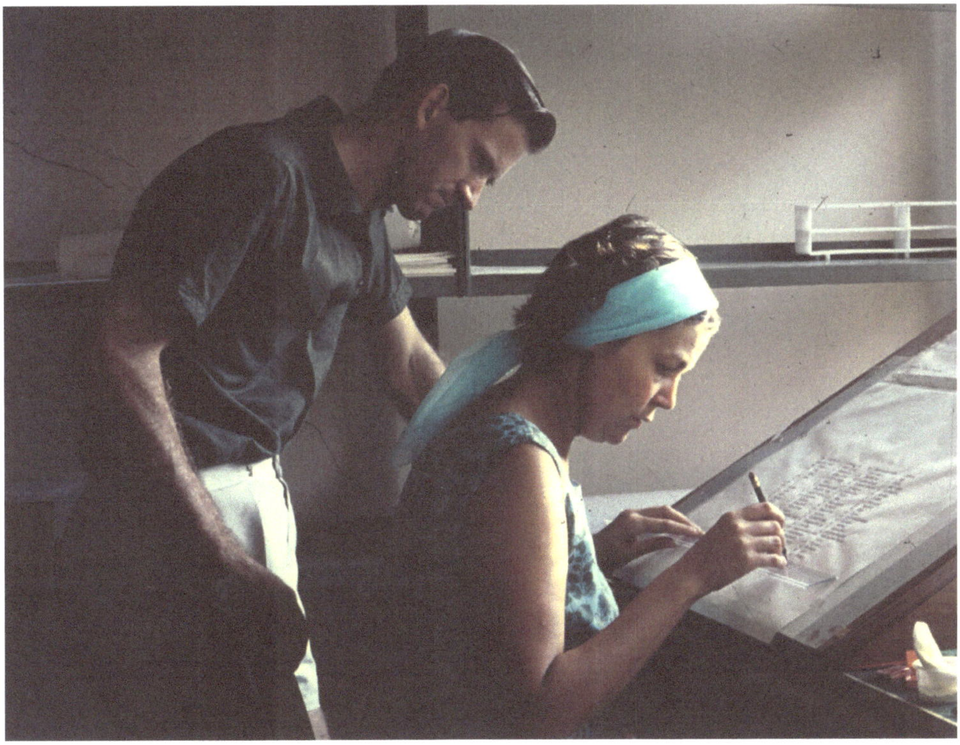

Figure 3a Peter and Sheila Waters pictured designing a restoration card.

LIB.	TITLE							VALUE
CAT. NO.	SYM-	ORIGINAL		REPAIR	25	Stabbed joint		COVERING
PHOT.	BOLS	STRUCTURE	1	As necessary		SEWING	50	Full
S	✉	COVER	2	General dry cleaning	26	Original holes	51	Limp
C	⊖	Full	3	General mending	27	Alum-tawed thongs	52	Yap edges
P	?	Stiff	4	Fold mending	28	Vellum thongs	53	Quarter and vellum tips
W	!!!	Limp	5	Reinforcement	29	Herring bone	54	Case
FC	~	Half	6	Swing plates	30	Double cords	55	Box
M	≈	Quarter	7	Do not trim edges	31	Single cords	56	Alum-tawed pig
Se.	≋	Alum-tawed pig		SIZING & ADHESIVES	32	Seaming cords	57	Vellum
For.	R	Vellum	8	Parchment size	33	Sawn-in cords	58	Calf
Fin.	R̲	Calf	9	Gelatine size	34	Linen tapes	59	Goat
WASHING YES NO	M	Goat	10	Starch paste	35	Linen braid	60	Native-dyed goat
OK	M̲	Sheep	11	Polyvinylacetate	36	Stab	61	Pulp
R/OK	X	Others	12	'Glutofix'	37	Link-sewing	62	Buckram
S	△	SEWING	13	Methyl cellulose (Tylose)	38	Linked overcasting	63	Reback
	⚠	Original	14	Soluble nylon	39	Machine	64	Refurbish
B	¢	Thongs		TYPES OF PAPER	40	Stabbed joints	65	Restore as original
	NT	Double cords	15	Handmade	41	As necessary		TITLING
T/A	⊄	Herring bone	16	Mouldmade		BOARDS	66	Original
D		Single cords	17	Machine-made	42	Laced-in	67	Manuscript
Ph Before After		Sawn-in cords	18	Japanese mending	43	'Split's	68	Tool
DEACIDIFICATION	—·	Tapes	19	Lens tissue	44	Original	69	Label
	·—	Laced-in	20	Heat-set tissue	45	Cased		TREAT WITH
		Two-on		ENDPAPERS		HEADBANDS	70	Saddle soap
REINFORCEMENT		HEADBANDS	21	'Italian'	46	Handsewn, laced-in	71	Potassium lactate
		Thread, laced-in	22	'Made'	47	Handsewn, thread	72	Paranitrophenol
		Thread	23	'Library'	48	Handsewn, silk	73	Lanolin & neatsfoot oil
?		Silk	24	Tipped	49	Stuck-on		
	OTHER INSTRUCTIONS							
!!!								

BIBL.	TITOLO							VALORE
SEGN.	SIM-	STRUTTURA		RIPARAZIONI	25	Impunturati sullo spigolo		COPERTA
FOTO.	BOLI	ORIGINALE	1	Secondo necessità		CUCITURA	50	Tutta
S	✉	COPERTA	2	Pulitura a secco	26	Tracce originali	51	Floscia
C	⊖	Tutta	3	Rattoppi generali	27	Linguette allumate	52	Labbro
P	?	Rigida	4	Rattoppi alla piega	28	Linguette di pergamena	53	¼ e punte cieche
W	!!!	Floscia	5	Rinforzo	29	Spina di pesce	54	A cartella
FC	~	½	6	Imbrachettare le tavole	30	Su nervo doppio	55	Custodia
M	≈	¼	7	Non tagliare i bordi	31	Su nervo	56	Porco allumato
Se.	≋	Porco allumato		SALDE E ADESIVI	32	Nervo fino	57	Pergamena
For.	R	Pergamena	8	Di pergamena	33	Nervo in traccia	58	Vitello
Fin.	R̲	Vitello	9	Di gelatina	34	Nastri di lino	59	Capra
LAVAGGIO SI NO	M	Capra	10	Pasta d'amido	35	Fettucce di lino	60	Capra tinta al naturale
OK	M̲	Agnello	11	Polivinilacetato	36	Impuntura	61	Cartoncino
R/OK	X	Altre	12	'Glutofix'	37	Cucitura intrecciata	62	'Buckram'
S	△	CUCITURA	13	Metil cellulosa ('Tylose')	38	Sopraggitto intrecciato	63	Rifare il dorso
	⚠	Originale	14	Nailon solubile	39	A macchina	64	Rinfrescare la legatura
B	¢	Linguette		TIPI DI CARTA	40	Impunturata sullo spigolo	65	Restaurare come l'originale
	NT	Su nervo doppio	15	A mano	41	Secondo necessità		TITOLI
T/A	⊄	Spina di pesce	16	A mano-macchina		QUADRANTI	66	Originale
D		Su nervo	17	A macchina	42	Incartonatura	67	Manoscritto
Ph Prima Dopo		Nervo in traccia	18	Giapponese per rattoppi	43	'Split'	68	Ferri
DEACIDIFICAZIONE	—·	Nastri	19	Da lenti	44	Originali	69	Cartellino
	·—	Intrecciata	20	Velina fissata a caldo	45	A cartella		TRATTAMENTO
		'Alternata'		FOGLI DI GUARDIA		CAPITELLI	70	Sapone da cuoio
RINFORZO		CAPITELLI	21	A brachetta	46	A mano intrecciati	71	Lattato di potassio
		Lino intrecciato	22	'Made'	47	A mano in lino	72	Paranitrofenolo
		Lino	23	'Biblioteca'	48	A mano in seta	73	Lanolina e olio 'neatsfoot'
?		Seta	24	Incollati a filo	49	A macchina		
	ALTRE ISTRUZIONI							
!!!								

Figure 3b examples of a condition card in English and Italian.

To add a personal note here, when I spent the month of May 1967 working with Peter in the library, one of my tasks was to design these cards, to draw the symbols and write all the wording in plain, legible calligraphy, camera-ready for printing by the thousand – one set in English, the other in Italian. Peter insisted that every book should be photographed before treatment and in less than two years over 50,000 volumes were photographed. He closes his paragraph about selection with a very important sentence: 'These symbols guard against the temptation to make the system more important than the books that are fed into it.'

Dry cleaning and pulling

> As much caked mud as possible is removed dry by flexing the edges of the leaves and the remainder is flaked off with spatulas and sharp blades. Then the book is collated and the sections are separated by careful cutting of the sewing from the spine. Covers, headbands, cords, threads, fragments etc. are placed in envelopes and catalogued. The sections are prepared for washing by interleaving with wet-strength paper at frequent intervals. Hand-colored prints are protected where necessary with a 3% solution of soluble nylon in alcohol.

Washing and drying

> Each leaf is supported on a floating wooden board and washed individually by soft brushing in warm water containing a saturated solution of Topane (2-hydroxy-diphenyl). This is followed by gentle squeezing of the whole book which is then pressed to remove excess liquid. Bleaching, de-acidification and re-sizing are done where necessary. At the power station the washing was done in rusty sinks covered in polyethylene sheeting. In the library, there are multiple stainless steel sink-units with forty compartments, and all are water-jacketed and thermostatically controlled. Random pH measurements are being taken with a Pye flat-head electrode pH meter before and after washing.

At the power station, sections were hung to dry on terylene lines, but in the library they are laid flat on racked trolleys in purpose-made, electrically warmed, automatically controlled drying cabinets. A final collation is made by a librarian and the book is then wrapped for storage.

I would like to add here that Sandy Cockerell, an engineer as well as a well-known private binder and restorer, was the prime helper and advisor in the design and fabricating of these large drying cabinets, working closely with Peter at the library during several visits. One of the many setbacks in the construction of these drying cabinets was finding that measurements had not been followed accurately and that the 16 trolleys were too large and had to be altered to fit.

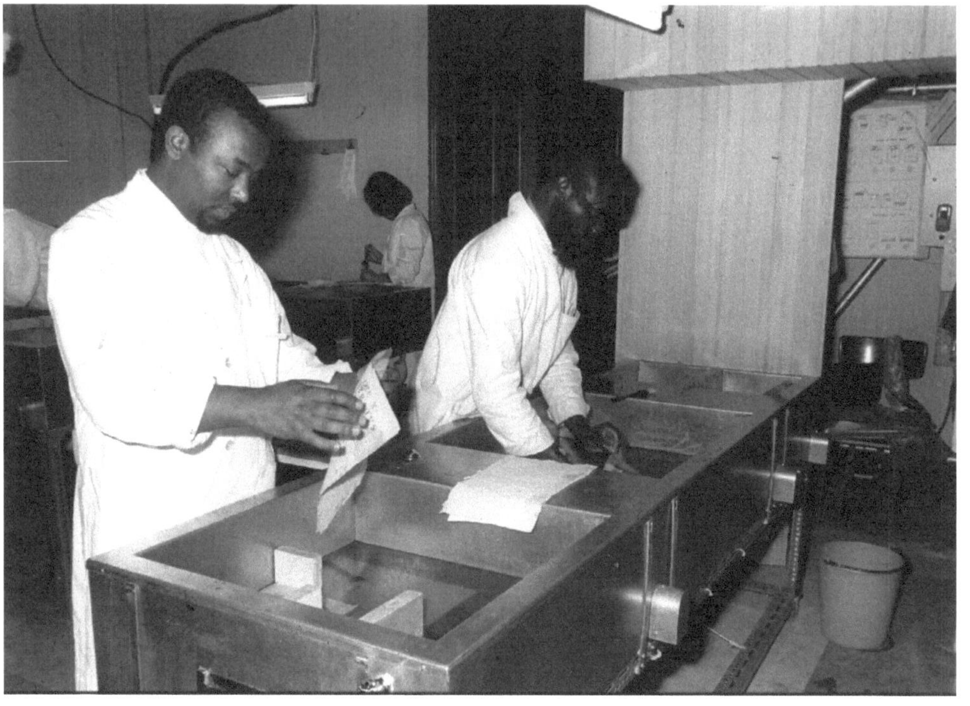

Figure 4 Washing at the BNCF.

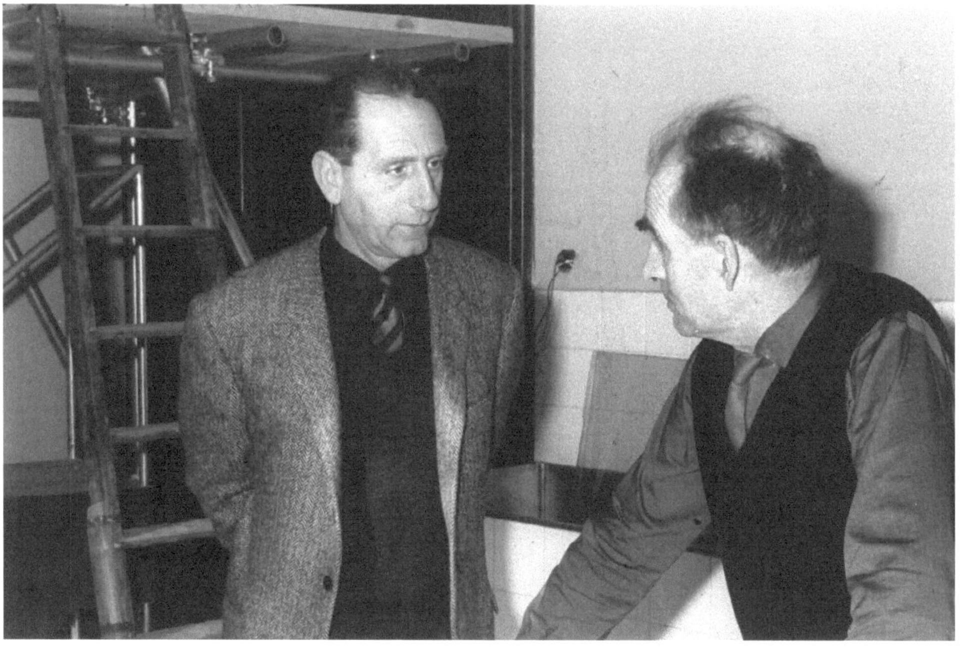

Figure 5 Professor Emanuele Casamassima, director of the BNCF (left), and Sandy Cockerell (right) with a trolley for a drying cabinet.

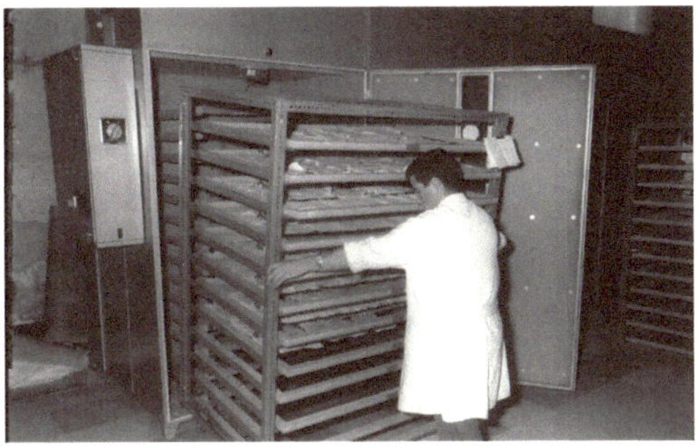

Figure 6 Trolley for the drying cart.

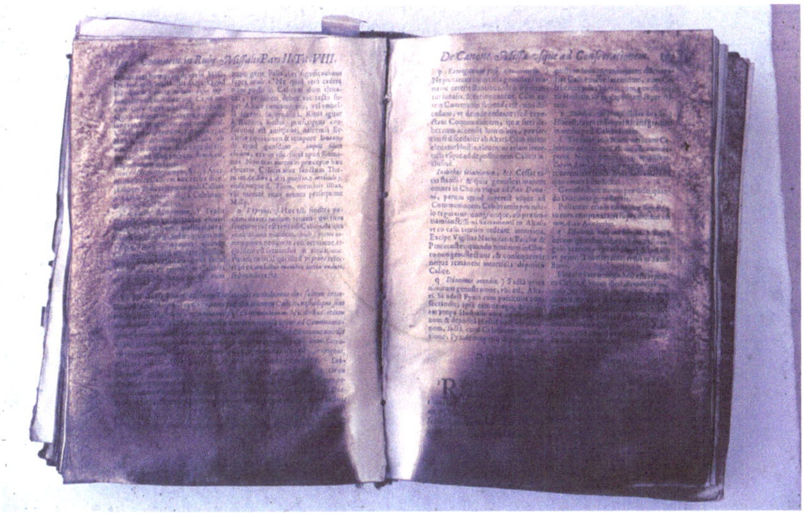

Figure 7 Gelatine stain.

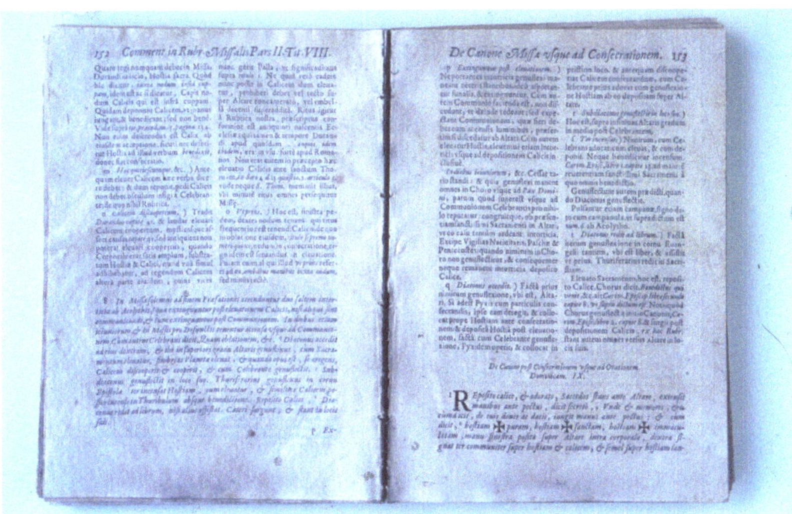

Figures 8 The pages illustrated in Figure 7 after treatment.

Mending

Peter's article continues with descriptions of the new mending and binding areas set up in the main reading room:

> Although the original intention was to send the prepared books to binderies throughout the world for mending and binding, by April 1967 it was apparent that this would be impracticable, mainly because of the wide diversity of standards and the impossibility of maintaining proper controls. It was therefore decided that mending, the most time-consuming operation of all, should be done in the library and that binding should follow. A mending department was designed for thirty workers (the maximum number for whom space and wages could be found), to be housed in the main reading room for at least a year or so. This was ready in August 1967, when the training of about five unskilled workers per week was undertaken until the room came into full operation. Mending methods had to be devised suitable for the scale of the operation and the condition of the paper. Long-fibered Japanese tissue paper is used to form patches, with adhesive designed to spread through to the area round the mend and remain flexible. A lens tissue is specially treated and used principally for the mending of tears. Many of the books have been reduced almost to single leaves and, on average, every section has at least two folds needing repair.

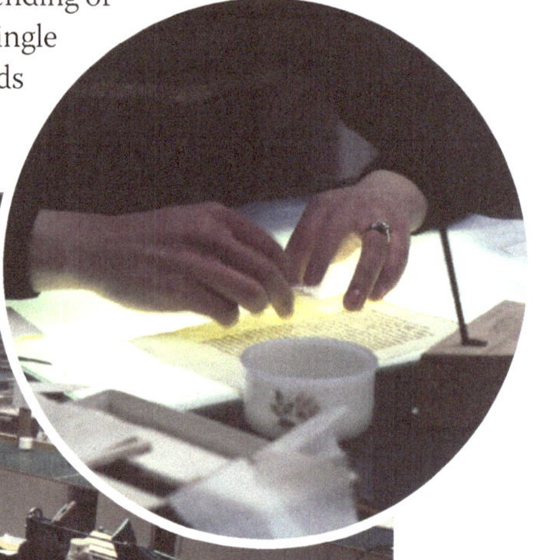

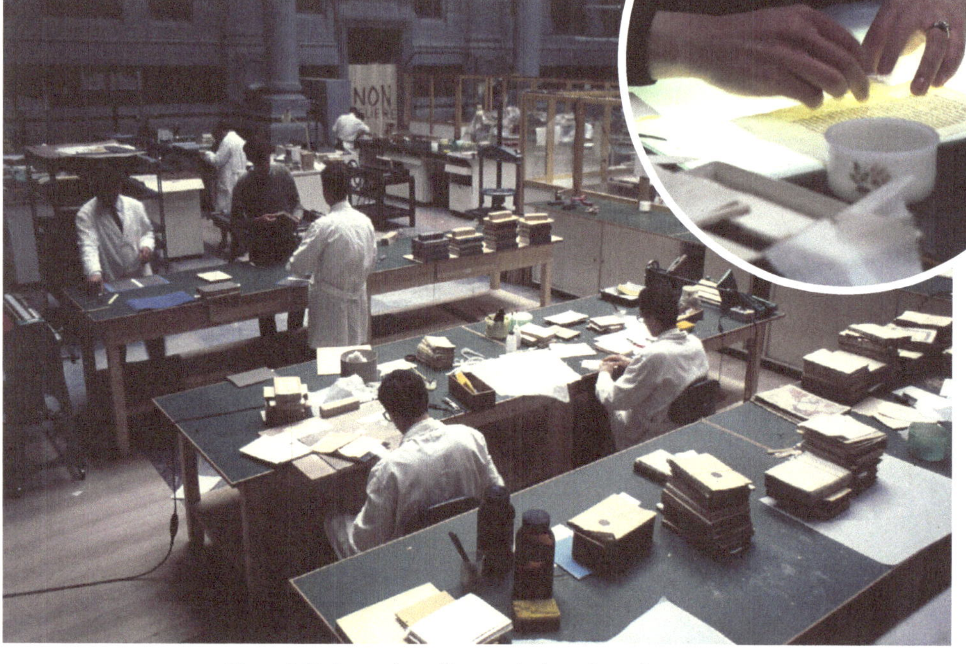

Figure 9 Binding and mending area in the main reading room.

One of my assignments during my stay in May was to help Peter design, then draw the plans and elevations for the 30 mending stations. Each had a small light box inserted into the desktop and each had plenty of storage space on either side. In August I had the immense thrill of seeing the desks come to life as they were brought in and set up.

Binding

Then Peter describes the new bindery, adjacent to the mending area:

> A bindery was started in September 1967. Many of the smaller books will be rebound in limp vellum. A study is being made of the early Italian limp vellum structures as so many of the library's books are bound in this way. When the structure is sound, it is a long-lasting binding, strong and pleasant to handle and fairly quick to make. A limp vellum style, unique to the library, is being evolved as, compared with the work of the 16th and 17th centuries, limp vellum binding of today has become decadent.

He added that it was intended to eventually extend the bindery to include library binding for the 350,000 post-1840 books.

Prints

Fine prints, many hand-colored, were affected and Peter then describes how they were treated:

> During the early summer of 1967 the Australian government provided money which was used to establish a print restoration department as an extension to the restoration system. The department was designed and set up by Mr. William Boustead, chief conservator to the National Art Gallery of New South Wales, Sydney, who also trained a nucleus of workers in print restoration. It is very well equipped, providing on-the-spot chemical analysis. One of the finest laboratories for print restoration existing in Europe today (in 1967 that is), it is now being looked after by a Ghanaian chemist Mr. Joe Nkrumah.

Conclusion

I hope you have been able to hear Peter through the sections of his article quoted above. He was able to leave the technical direction of the system in the capable hands of Tony

Cains, whose tenure was supported financially by the American fund CRIA, for the following three years. During that time, mostly back in England, resuming his part-time teaching at the Royal College of Art, Peter joined with a co-director, James Lewis, in a research program to support the work of the BNCF. This program, overseen by the Imperial College of Science and Technology and the Royal College of Art, was funded by the US Council on Library Resources. In Rome, the Istituto di Patologia del Libro also continued to support research.

The student labor force at the BNCF was gradually replaced by about 100 Italian workers. With that number of people, it was estimated that the restoration would take about 25 years, but the workforce has dwindled through lack of financial resources so the work still continues. A dream of eventually setting up an international training centre at the library has never materialized because of insufficient funding and too few expert restorers available to continue training. But the legacies from the BNCF experience gave birth to a new profession – that of the library conservator. It is now possible to study for not only a master's degree but also a PhD in this subject.

In 1971, Peter took up the position of restoration officer at the Library of Congress – later to be called chief of conservation. He was given the task of designing and setting up a large comprehensive department from scratch, and from 1969 to 1971 he 'commuted' ten times to carry out this task. This position was offered to him as a direct result of the work he had done in Florence. Because of the many lessons he had learned at the BNCF about dealing with damaged library material on a vast scale, he was able to persuade the curators of the Library of Congress to think more widely in terms of what Peter called 'phased conservation' instead of merely sending single items for repair to the library's bindery. Since then, conservation laboratories have been set up throughout the world, many of them directed and staffed by conservators trained at the Library of Congress.

The aim of this paper has been to set the scene of those early months in the BNCF. Tragedies and disasters often bring out the best in people involved, and the great Florence flood certainly did that. So much good has come from its aftermath in the world of book and paper conservation, and many of the skilled people who labored there have since taught and influenced countless others. I feel blessed and honored to have played a small part in the effort myself, as a support to my late husband.

The work of the restoration centre in the Biblioteca Nazionale Centrale di Firenze 1967–1971

Anthony Cains

This paper shows some black and white photographs of the early period in the immediate aftermath of the flood and colour images illustrating the restoration system that was ultimately developed at the Biblioteca Nazionale Centrale di Firenze (BNCF).

We were met at the station by Dr Capara from Bologna University, to whom we first volunteered our services, and eventually we became accredited by the British Library through Howard Nixon and the Art and Archive Rescue Fund. I don't remember the processes that were involved, but I know that many people volunteered, sending telegrams offering their services. I was asked to volunteer by my friend John Corderoy of Camberwell School of Arts and Crafts prompted by his star pupil Sally Lou Smith. So at the end of November we found ourselves in Florence – the first group, Peter Waters, Dorothy Cumpstey and me, sent by the British Fund. Dorothy's accommodation was at a pensione in Piazza Santissima Annunziata and we lads were billeted at a monastery nearby. (After a few days Peter was promoted to digs at Villa I Tatti following our contact with Paul Banks of the Committee to Rescue Italian Art (CRIA) fund.) My cell was very damp and the blankets exceedingly heavy but fortunately I had brought my army surplus rucksack and sleeping bag, and that gave me some comfort along with a heavy leather trench coat – I had anticipated such trouble.

The illustrations start with a picture of Peter and Dorothy wending their way to the BNCF, with me in train with a camera (Fig. 1). On arrival we see soldiers and students (Fig. 2) – the 'mud angels' – taking a break. Inside we have a picture of Signora Casamassima (the director's wife), and Alberto Cotogni, one of the many dedicated librarians, preparing food for the volunteers (Fig. 3). Following a brief meeting with the director, Professor Emanuele Casamassima, we were taken to the Forte del Belvedere to meet Dr Luigi Crocetti and Dr Carla Bonanni supervising the identification and stacking of dried books being returned from the drying kilns. There, we also met up with, among others, the CRIA representatives – Paul Banks of the Newberry Library and Harold Trebolet of the Lakeside Press (Fig. 4). We were interrogated by Paul Banks as to our veracity/authenticity, however once the name of Peter's business partner, Roger Powell, was mentioned, he seemed to be convinced! Figure 5 shows the dried books being lifted above the parapet of the Forte di Belvedere.

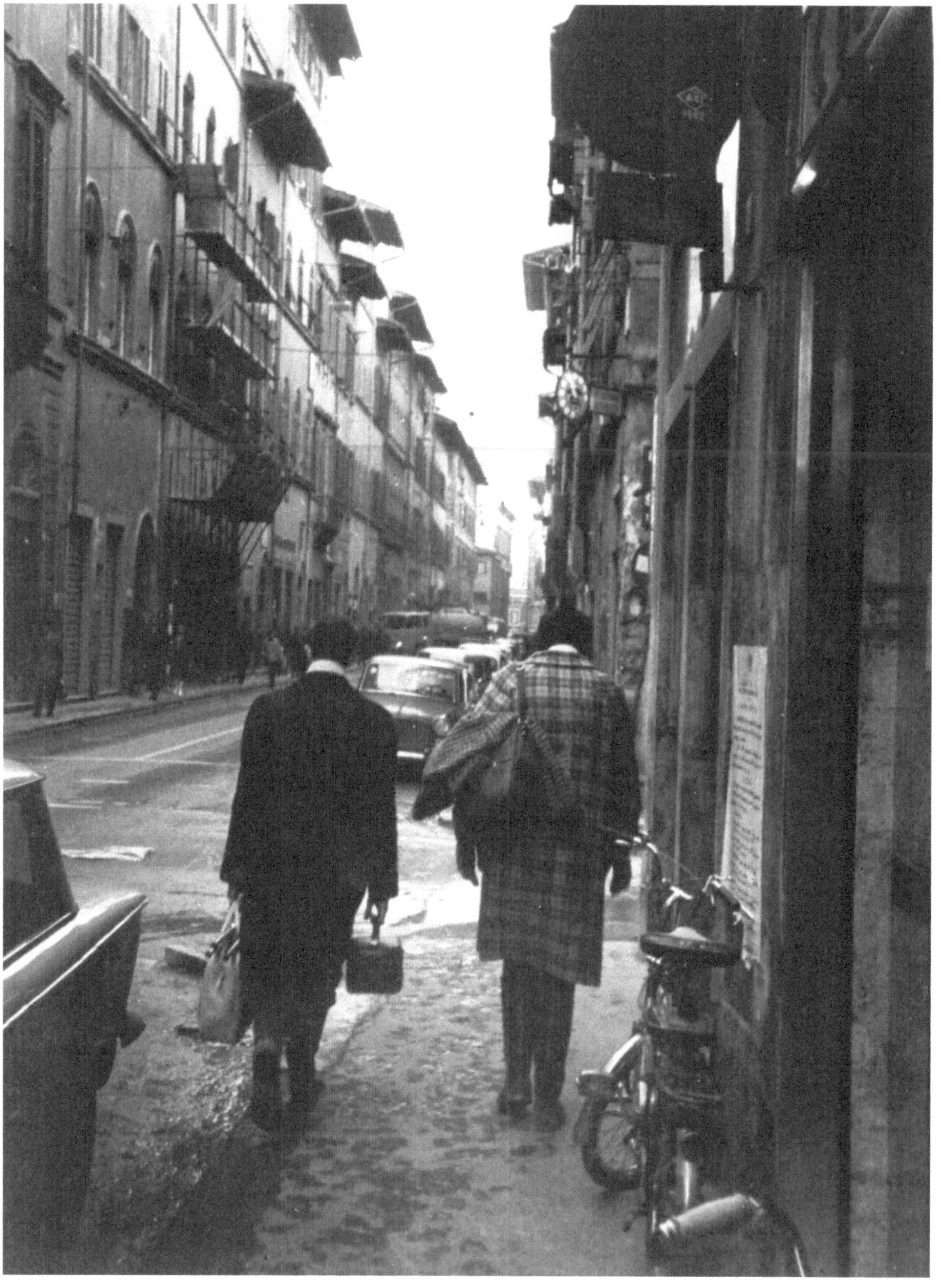

Figure 1 Peter Waters and Dorothy Cumpstey en route to the Biblioteca Nazionale Centrale di Firenze (BNCF).

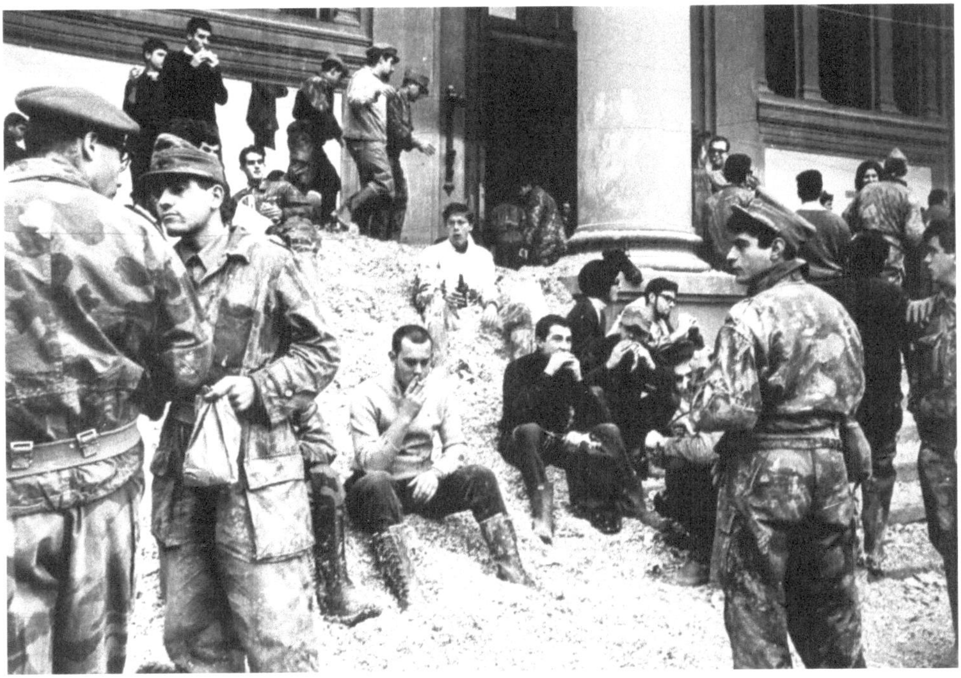

Figure 2 Soldiers and students – the 'mud angels' – at rest.

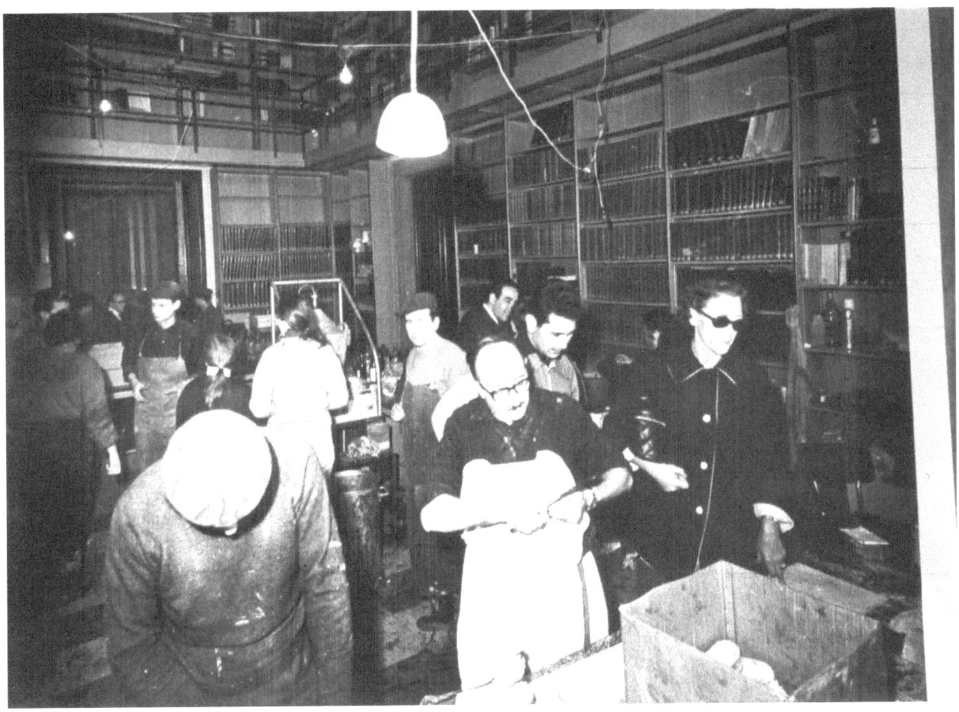

Figure 3 Signora Casamassima (the director's wife) and Alberto Cotogni, one of the librarians at the BNCF preparing food for volunteers.

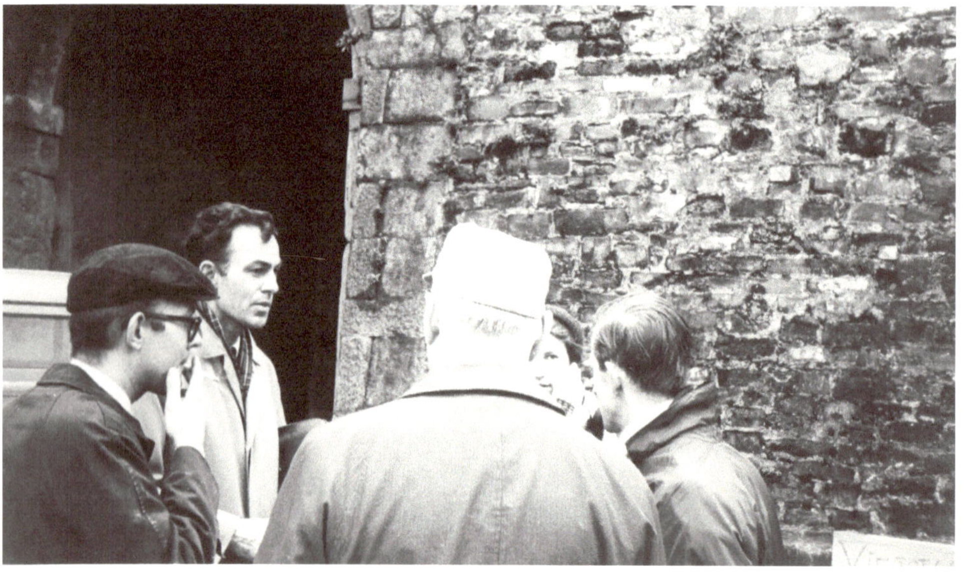

Figure 4 Meeting with the CRIA representatives, Paul Banks and Harold Trebolet.

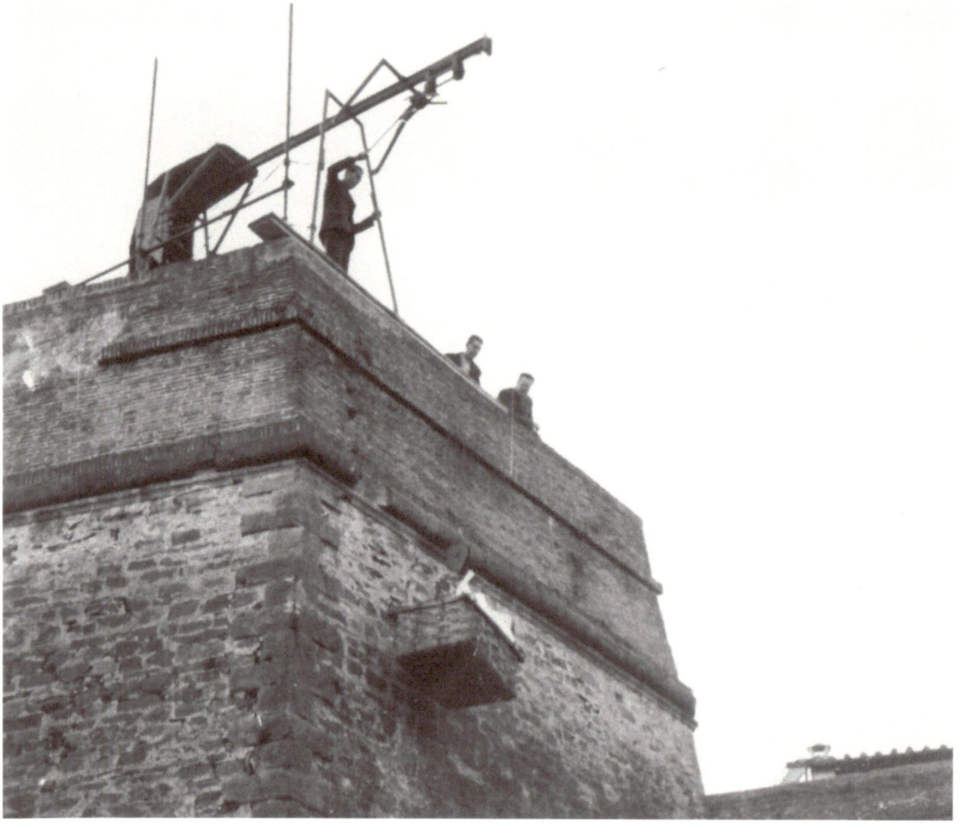

Figure 5 Dried books being lifted over the parapet of the Forte di Belvedere.

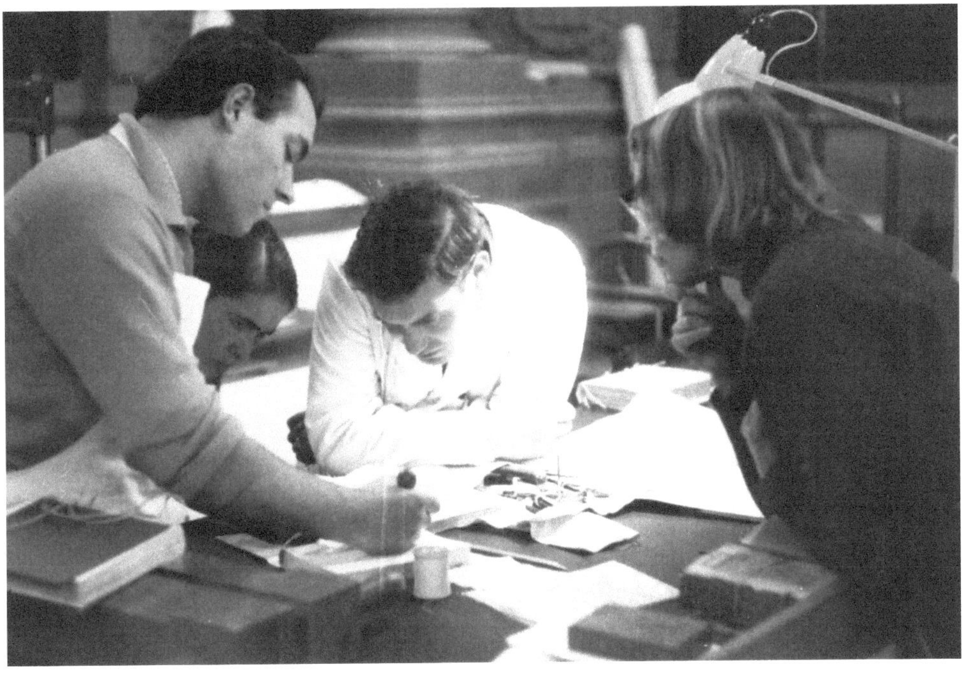

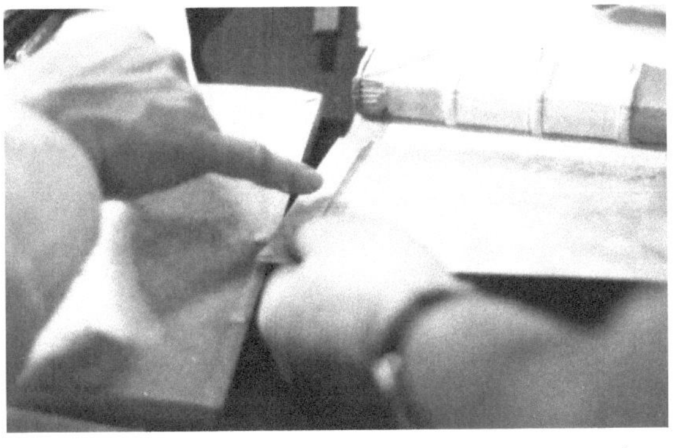

Figures 6 and 7 Christopher Clarkson making a technical point to Don Etherington, Sally Lou Smith and Barbara Guiffrida.

The following figures illustrate the establishment of the centre and the methods adopted or invented. Figures 6 and 7 show Christopher Clarkson making a technical point to a group of colleagues: Don Etherington, Sally Lou Smith and Barbara Guiffrida. Frank Tushingham of the British Library, who was in charge of the Barrow deacidification and buffering process and who along with his colleague George Jolly did much to teach and supervise the young Italian staff, can be seen in Figure 8. Figure 9 shows Cornelius Messerschmidt (Swiss fund) and Desmond Shaw, assistant to Sandy Cockerell, playing tug-of-war. Our Danish friends (from the Nordic Centre), Henning Madson and Arne Pedersen are illustrated in Figure 10, followed by a picture of Dag-Ernst Petersen working on the *Teuerdank* (Fig. 11), some time in 1968. Mammo Bogale from Ethiopia, who carried out sterling work ordering materials and keeping the store tidy, can be seen in Figure 12. Finally as a representative of the many American conservators who worked in the system, Norvell Jones of the Library of Congress is shown in Figure 13.

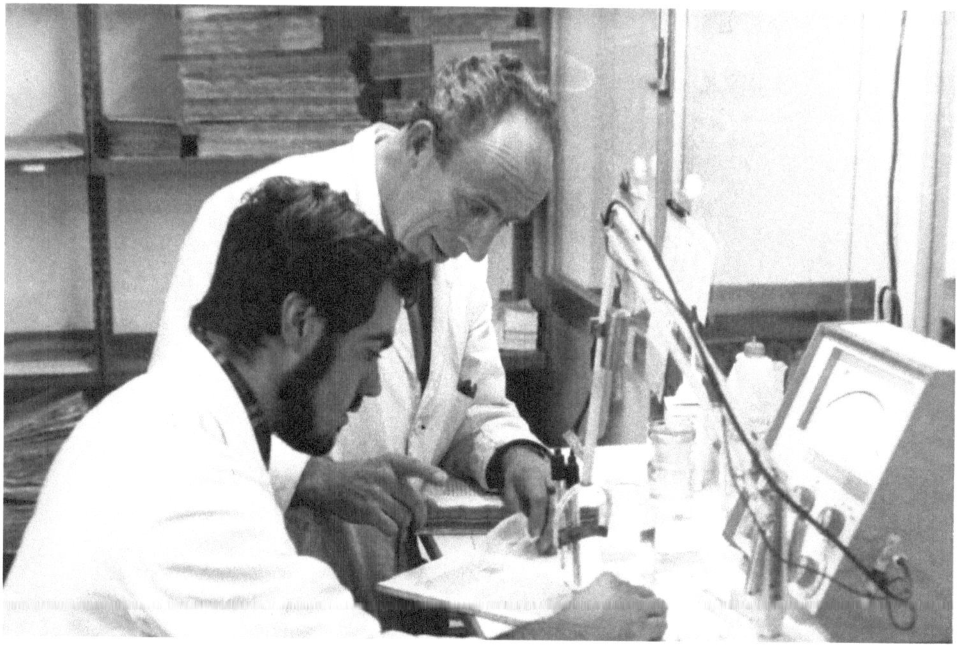

Figure 8 Frank Tushingham of the British Library supervising a young Italian.

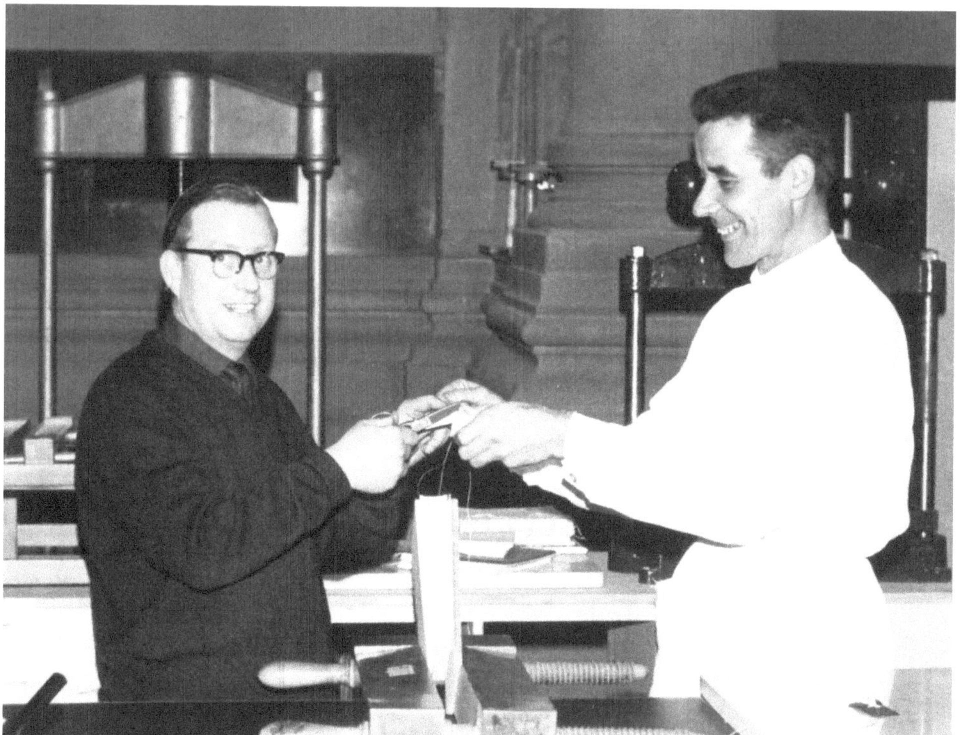

Figure 9 Cornelius Messerschmidt and Desmond Shaw, assistant to Sandy Cockerell.

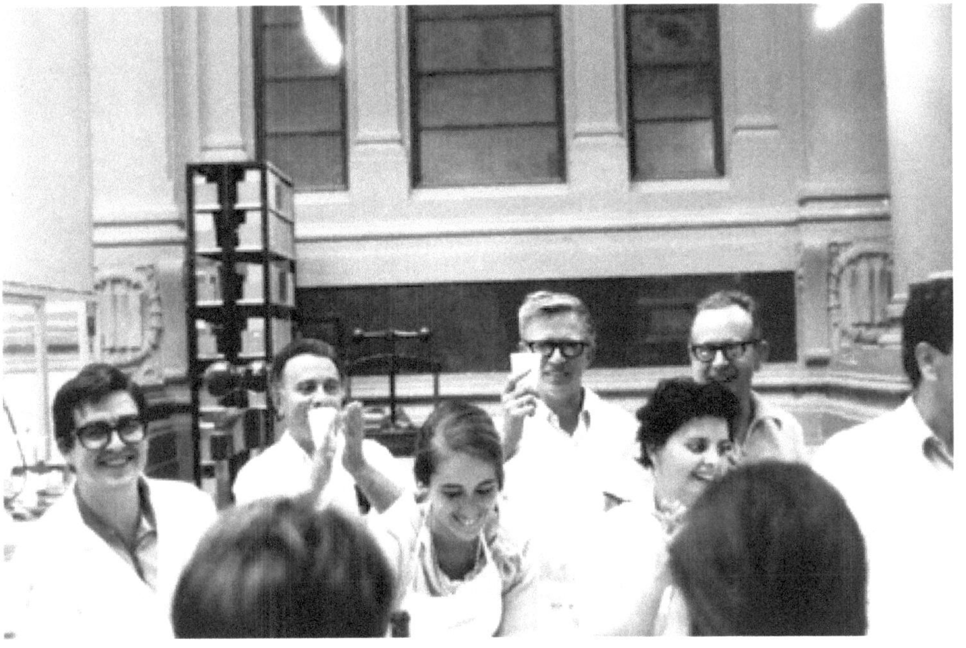

Figure 10 Danish friends including Henning Madson and Arne Pederson (to the right of the picture).

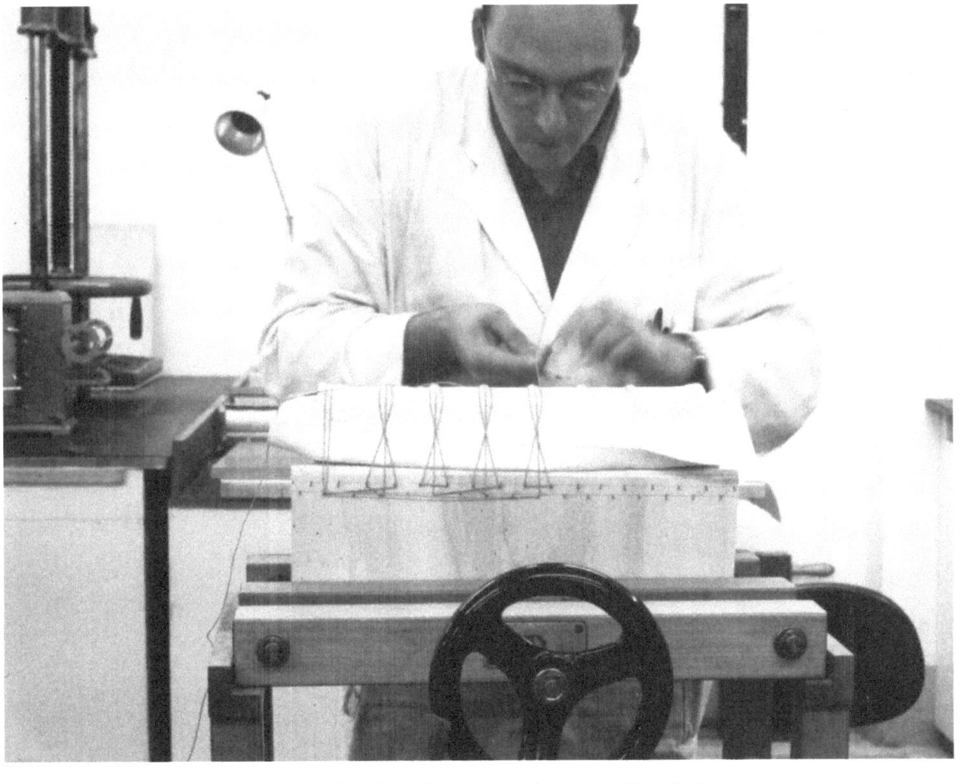

Figure 11 Dag-Ernst Petersen working on the *Teuerdank*.

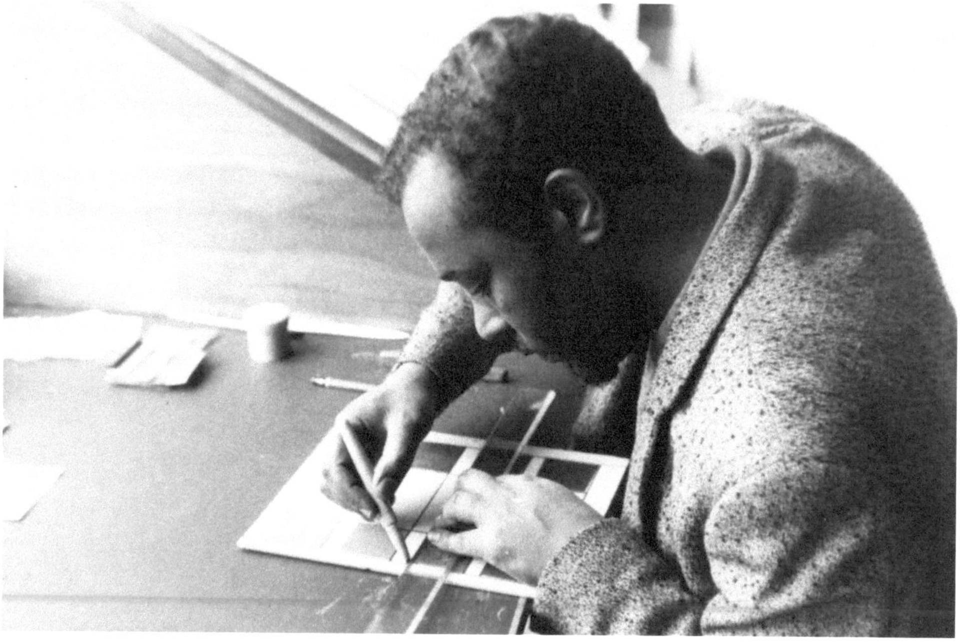

Figure 12 Mammo Bogale from Ethiopia.

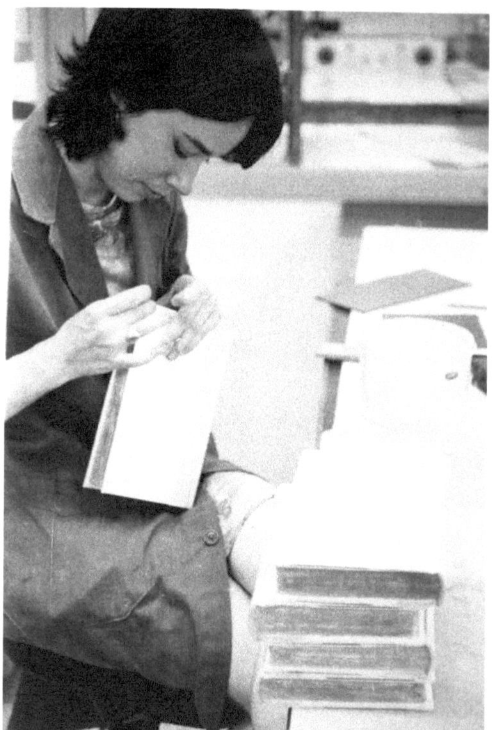

Figure 13 Norvell Jones from the Library of Congress.

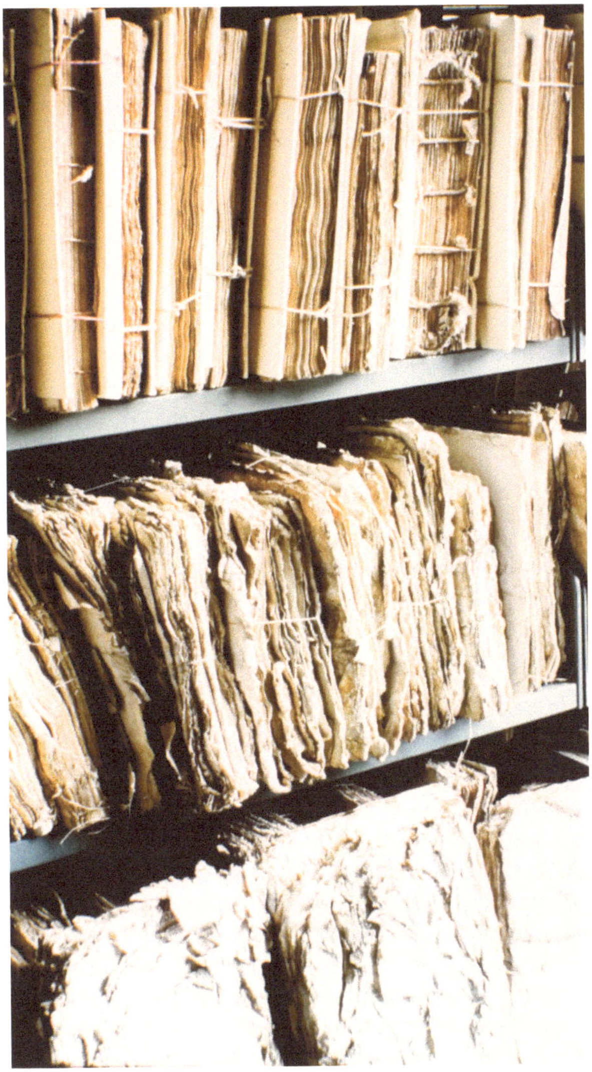
Figure 14 Dried text minus their covers.

During the period immediately following the flood, the mud-covered books were sent to several centres for treatment. According to Professor Casamassima, about 3000 tons of flood-damaged wet books were sent to be dried in many centres in Italy, taking upwards of 530 lorry loads. They were sent to the tobacco-drying kilns of Bologna, Citta di Castello, San Guistino Umbro; the brick kilns of Borgo san Lorenzo and the wood-drying kilns of Valdarno, Prato; a food-drying plant at Lucca; and to the University of Bari and the Ministry of Public Instruction, Rome. The quality of the treatment varied from centre to centre. Figure 14 illustrates some dried texts minus their covers. The higher boarded and neatly tied books on the shelf look clean and well processed – they came from the Bologna authorities, probably under Dr Capara's supervision. The books below are all rather tattered and arrived from somewhere south of Florence. Some of the detached covers being dried can be seen in Figure 15. The one shown in Figure 16 for example (a sixteenth-century gold and blind tooled calfskin cover) could easily be replaced onto its text block, but regretfully, many of the original components have been discarded or lost – including the head and tail bands, the paste boards, and perhaps the end-papers. The next example is of a gold-tooled vegetable-tanned goatskin binding encrusted with sawdust and mud (Fig. 17). After cleaning (Fig. 18) there is a significant improvement but inside, the text has massive tannin stains from the cover itself (Fig. 19). This would have been dealt with in the Restauro di Stampe under Joe Nkrumah's supervision.

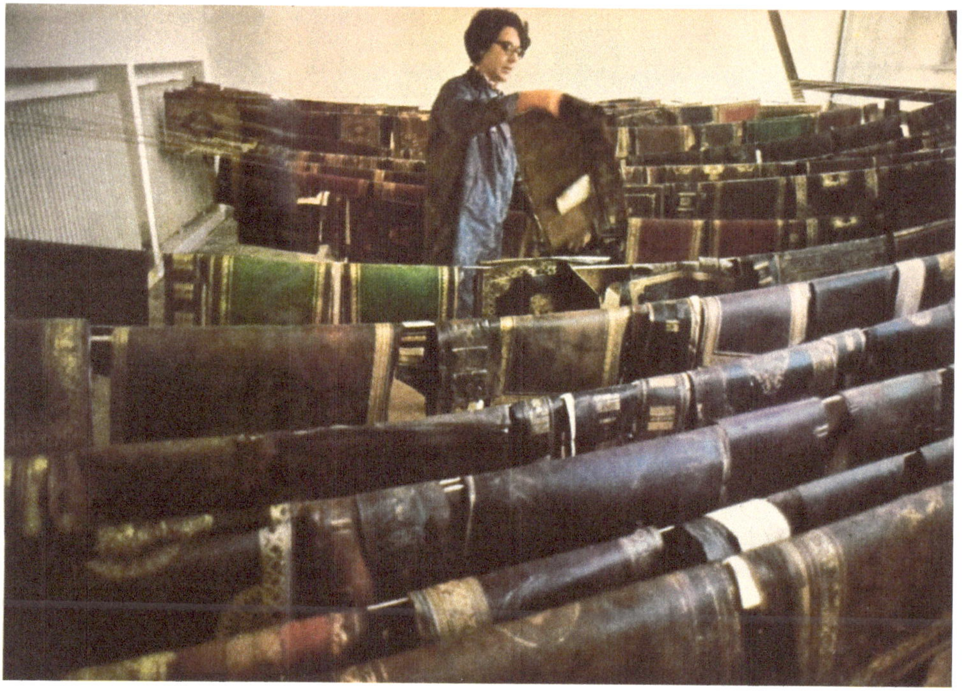

Figure 15 This illustration shows Dr Cochetti hanging detached gold-tooled leather covers to dry (*Sunday Times Magazine* 11 December 1966: 26 and 27).

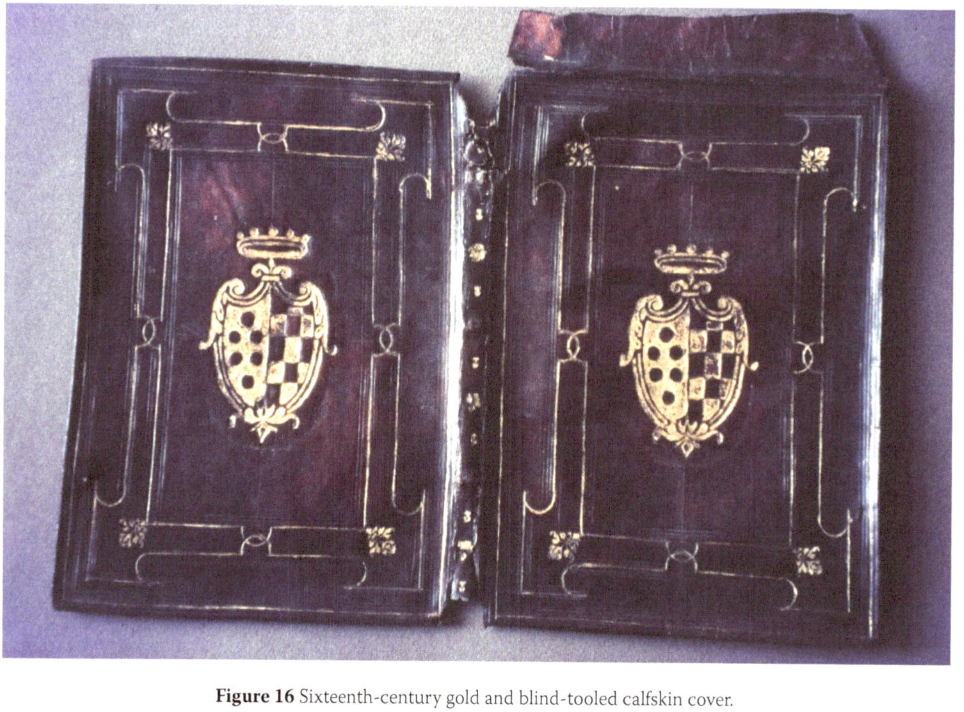

Figure 16 Sixteenth-century gold and blind-tooled calfskin cover.

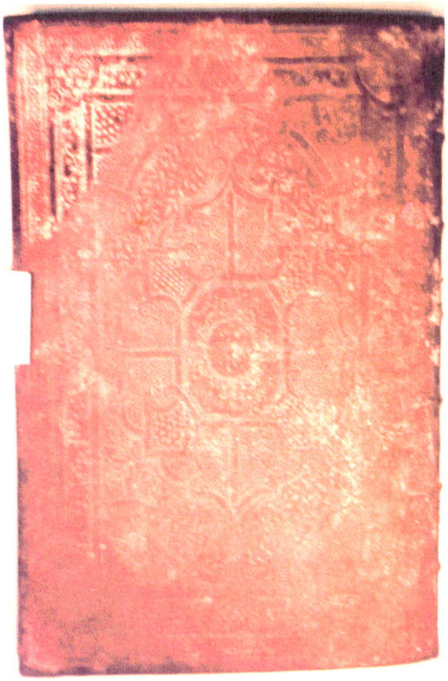

Figure 17 Gold-tooled vegetable-tanned goatskin encrusted with sawdust and mud.

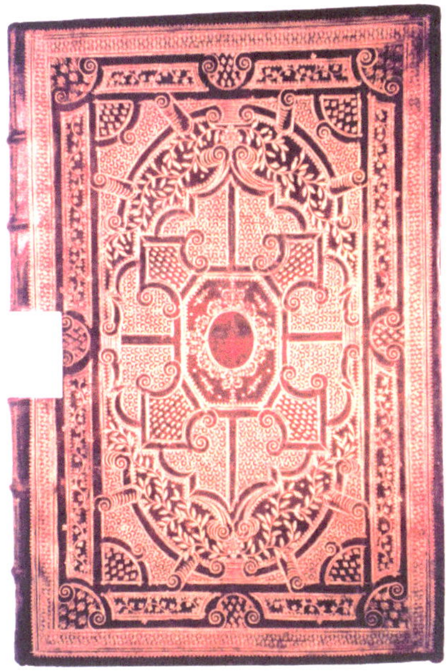

Figure 18 Gold-tooled vegetable-tanned goatskin after cleaning.

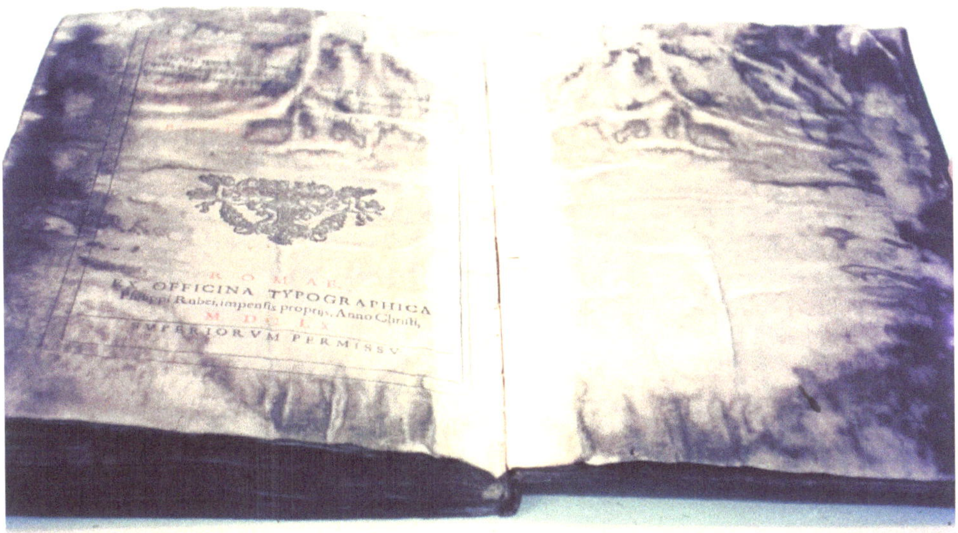

Figure 19 Text block with massive tannin stains from the cover itself (gold-tooled vegetable-tanned goatskin).

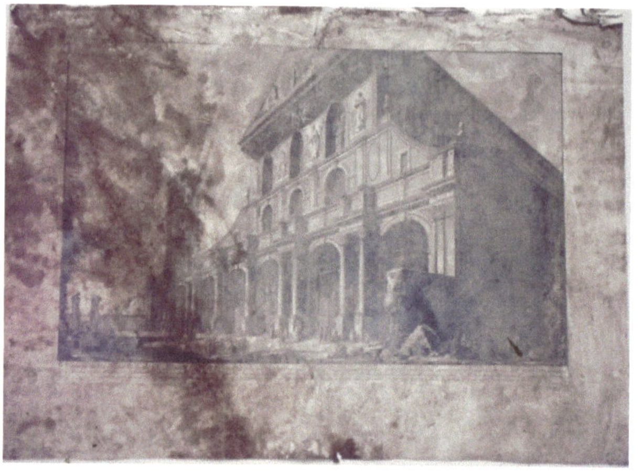

Figure 20 Piranesi print on waterleaf paper impregnated with fuel oil and mud.

Another example of damage is a Piranesi print on Waterleaf paper impregnated with fuel, oil and mud (Fig. 20). This sort of damage was treated with Fuller's earth and various solvents but I am not certain whether this treatment was successful. Figure 21 shows the filter containing a fungicide, Topane WS, an orthophenylphenol, specified by the British Museum Research Laboratory under Anthony Werner, through which water was passed into a large tank that supplied the washing sinks (Fig. 22). These figures illustrate the construction of the washing department (*lavaggio*). The drying racks and cabinets (Figs 23–26) were made by local engineers (Sandy Cockerell participated in the design process). The stainless steel water-jacketed sink and a book support (which I sketched out together with Philip Smith) are illustrated in Figures 27 and 28. Figure 29 is an enlargement of the selection symbols designed at the Forte di Belvedere. This is the original document and additions in various hands can be seen including mine and Dorothy Cumpstey's (this was part of the triage process for selecting books for the Centro Termica to be disbound and washed).

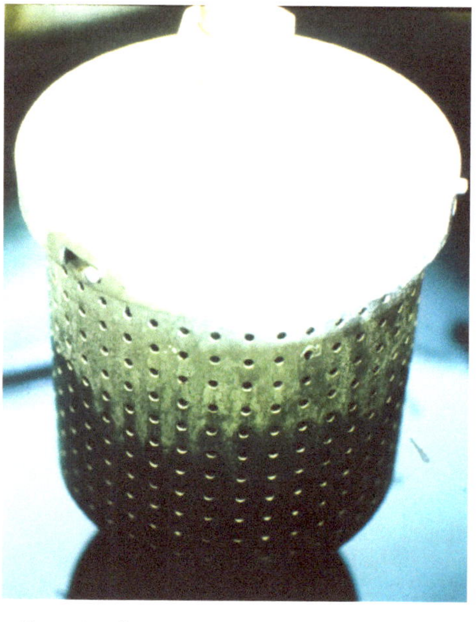

Figure 21 Filter containing a fungicide (Topane WS).

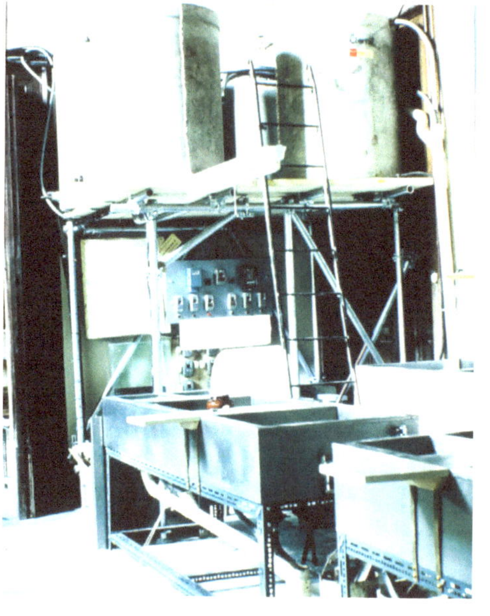

Figure 22 Large tanks and washing sinks.

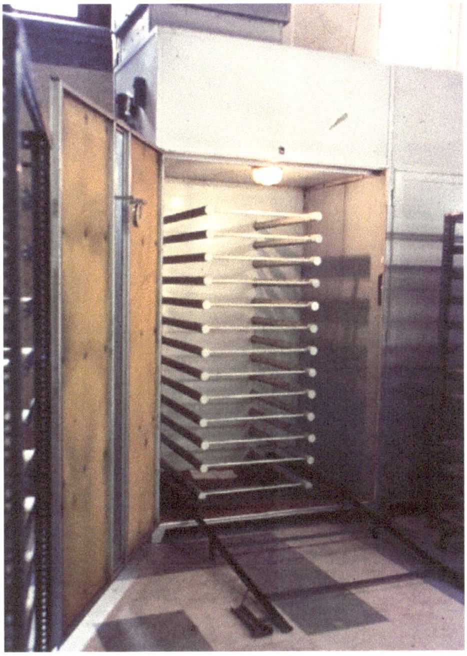

Figure 23 Drying racks and cabinets.

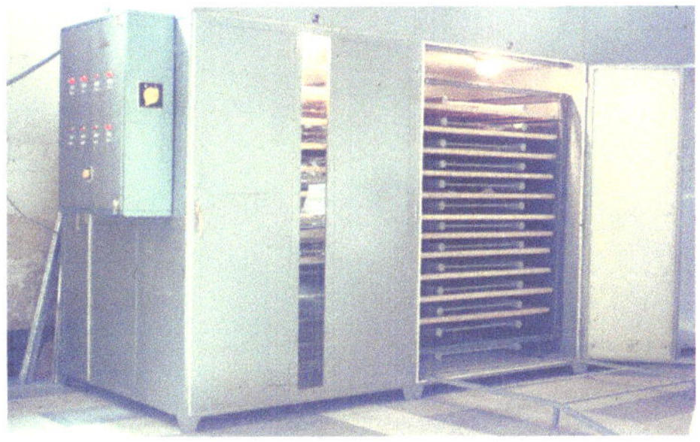

Figure 24 Italian-made drying cabinets.

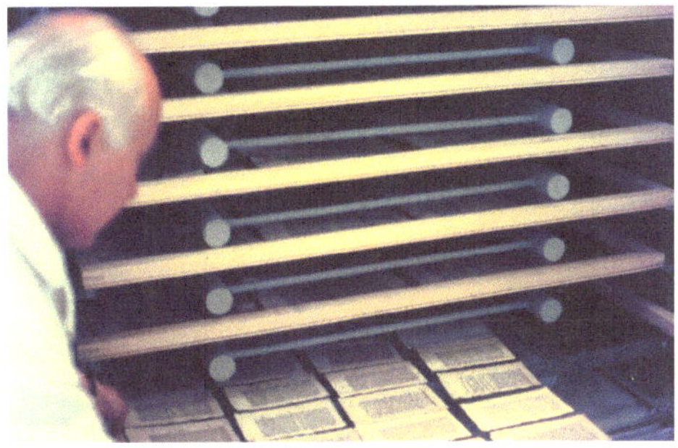

Figure 25 Washed paper leaves loaded on screens and slid into wheeled racks. Note the hot air pipes between the screens.

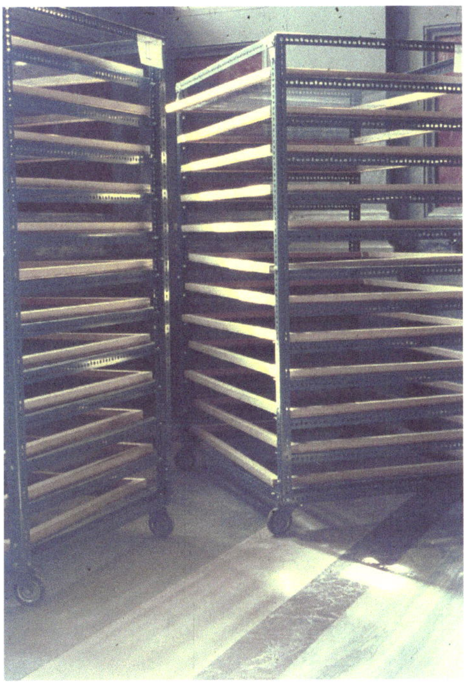

Figure 26 Wheeled racks of empty screens.

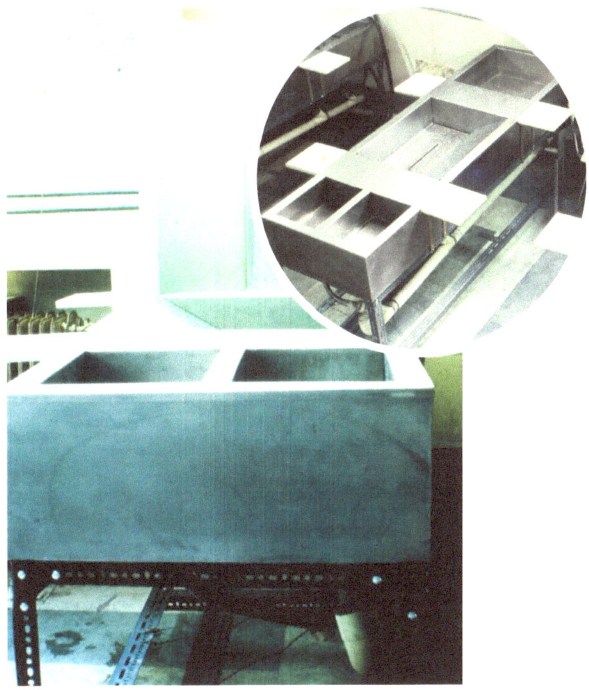

Figures 27 and 28 Stainless steel water-jacketed sink with book support.

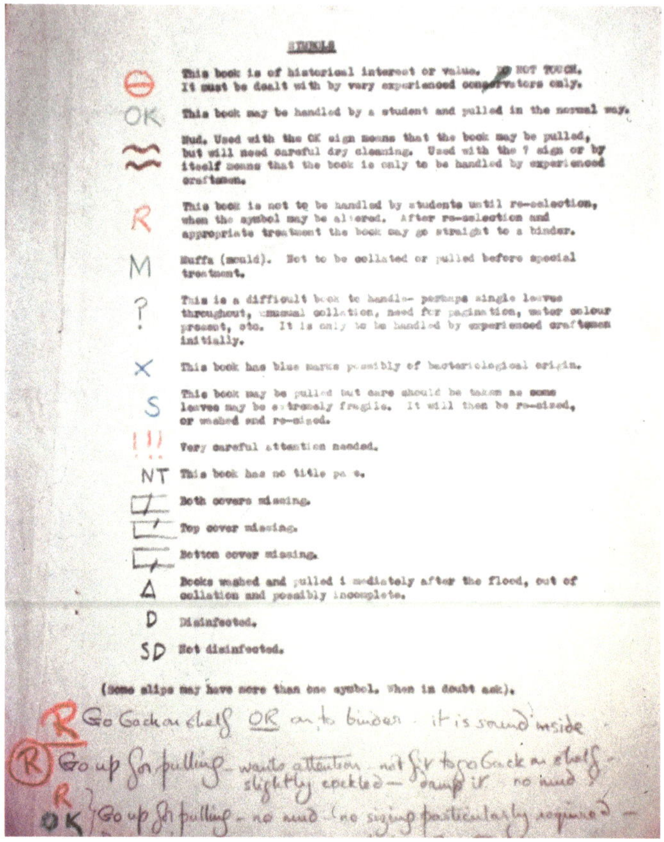

SYMBOLS

⊖ This book is of historical interest or value. DO NOT TOUCH. It must be dealt with by very experienced conservators only.

OK This book may be handled by a student and pulled in the normal way.

≈ Mud. Used with the OK sign means that the book may be pulled, but will need careful dry cleaning. Used with the ? sign or by itself means that the book is only to be handled by experienced craftsmen.

R This book is not to be handled by students until re-selection, when the symbol may be altered. After re-selection and appropriate treatment the book may go straight to a binder.

M Muffa (mould). Not to be collated or pulled before special treatment.

? This is a difficult book to handle– perhaps single leaves throughout, unusual collation, need for pagination, water colour present, etc. It is only to be handled by experienced craftsmen initially.

X This book has blue marks possibly of bacteriological origin.

S This book may be pulled but care should be taken as some leaves may be extremely fragile. It will then be re-sized, or washed and re-sized.

!!! Very careful attention needed.

NT This book has no title page.

Both covers missing.

Top cover missing.

Bottom cover missing.

Δ Books washed and pulled immediately after the flood, out of collation and possibly incomplete.

D Disinfected.

SD Not disinfected.

(Some slips may have more than one symbol. When in doubt ask).

R Go back on shelf OR on to binder – it is sound inside

(R) Go up for pulling – wants attention – not fit to go back on shelf – slightly cockled – damp it – no mud

R OK Go up for pulling – no mud – (no sizing particularly required) –

Figure 29 Selection symbols.

The processes at the BNCF

This material had probably been fumigated in an ethylene oxide chamber paid for by CRIA and specified by the Patologia del Libro. Local mould debris was vacuumed (Fig. 30) and then the text checked before pulling (Fig. 31). The pulling department in the library, initially organised by George Kirkpatrick, was where the books were disbound and mud-blocked leaves separated (Fig. 32). They were dry cleaned, i.e. plates of mud and particles were peeled off by chipping away with a spatula (Fig. 33). This was followed by another collation and cutting the interleaving of wet-strength filter paper supports (Fig. 34) before going to the washing department. Due to concerns about the heat from the vats, thermal insulation panels were fitted and Philip's support discarded in favour of a floating board of marine plywood (Fig. 35). The leaves were supported with interleaving and the mud streaks brushed away (Fig. 36). Figures 37 and 38 show Italian staff in the laboratory preparing the Barrow buffering solution. I recall that the tap water was quite alkaline. Deionised water was only available in small quantities and used for the processing of microfilm.

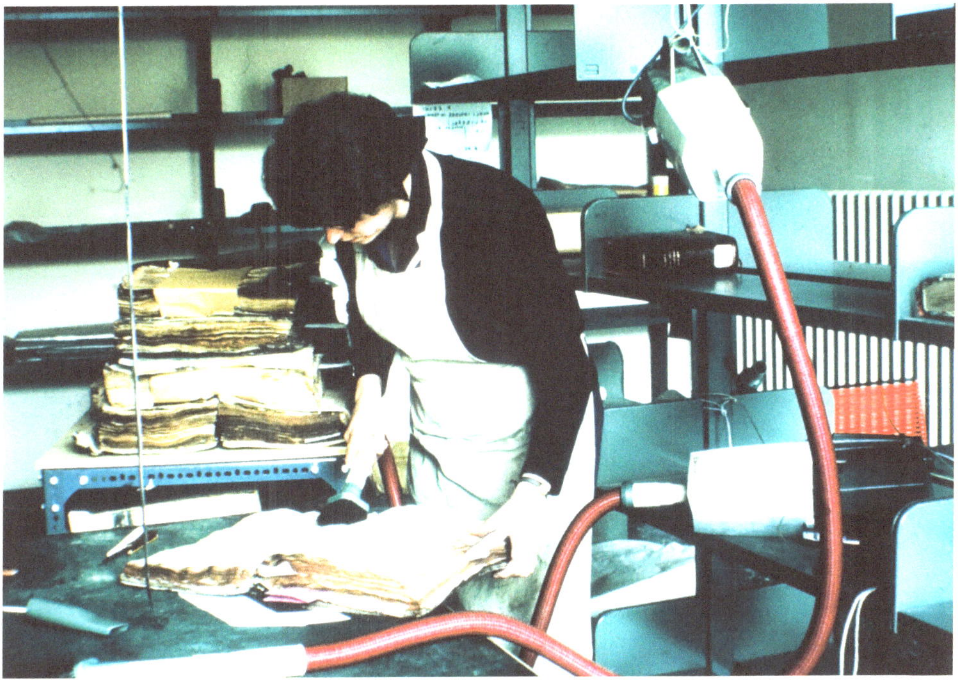

Figure 30 Local mould debris being vacuumed.

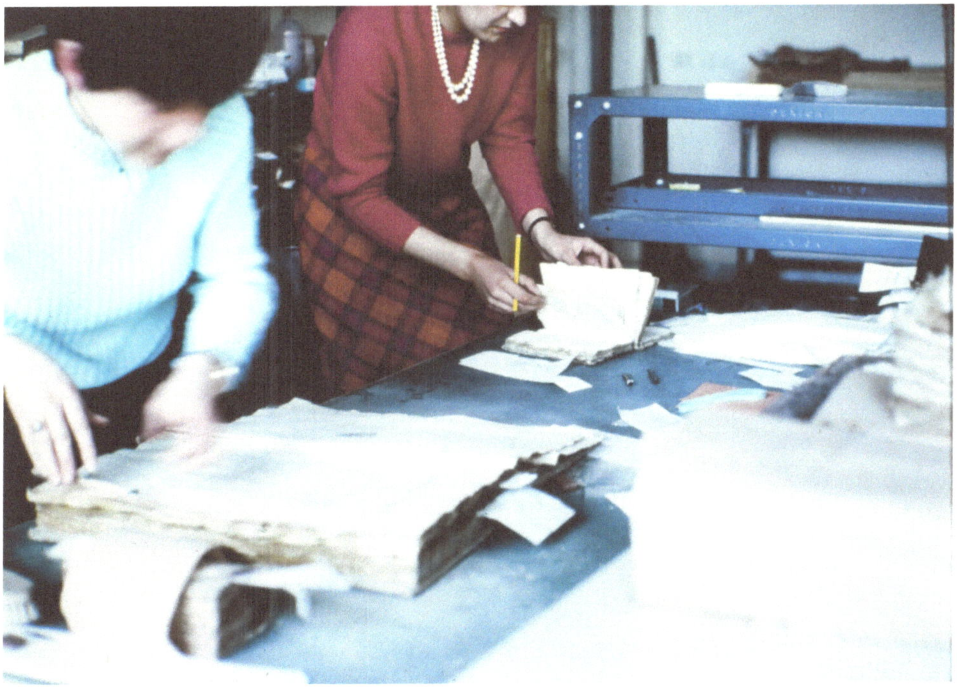

Figure 31 Text check before pulling.

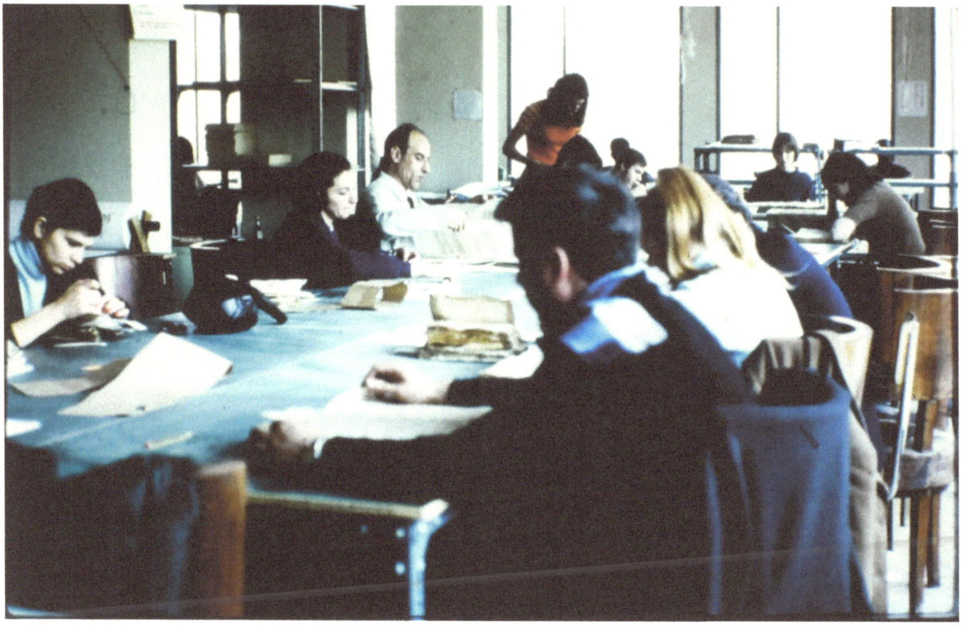

Figure 32 Pulling department.

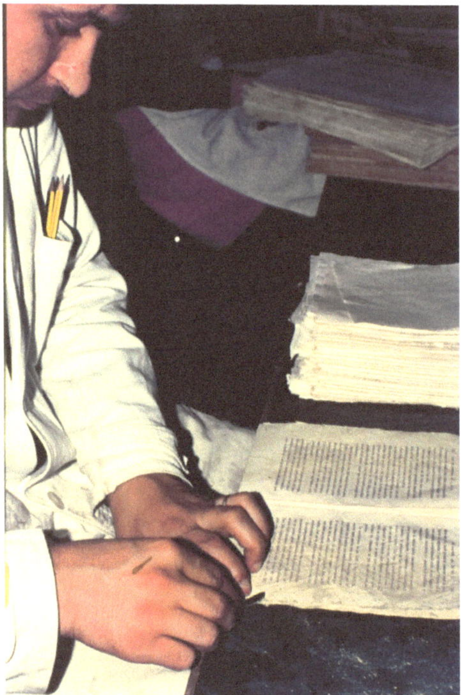

Figure 33 Chipping away with a spatula.

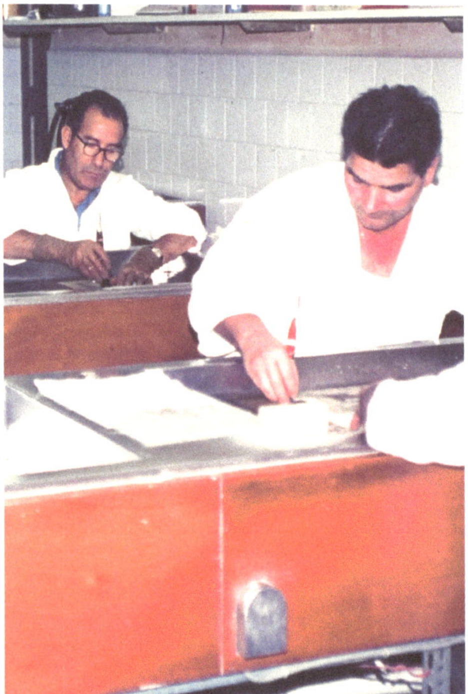

Figure 34 Collation and cutting the interleaving of wet-strength filter paper supports.

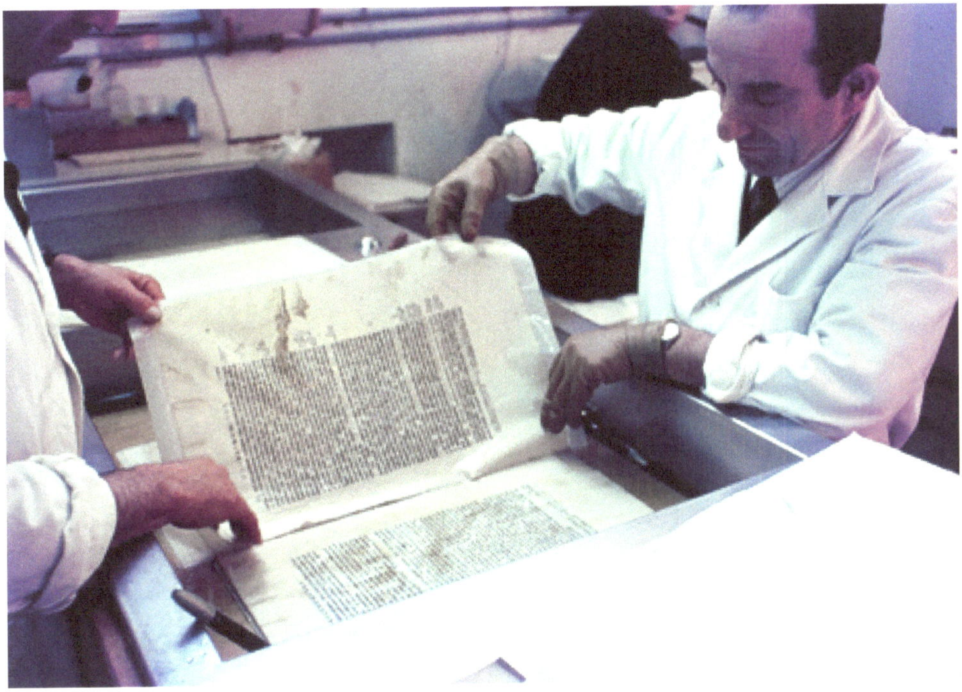

Figure 35 Washing department: floating board of marine plywood.

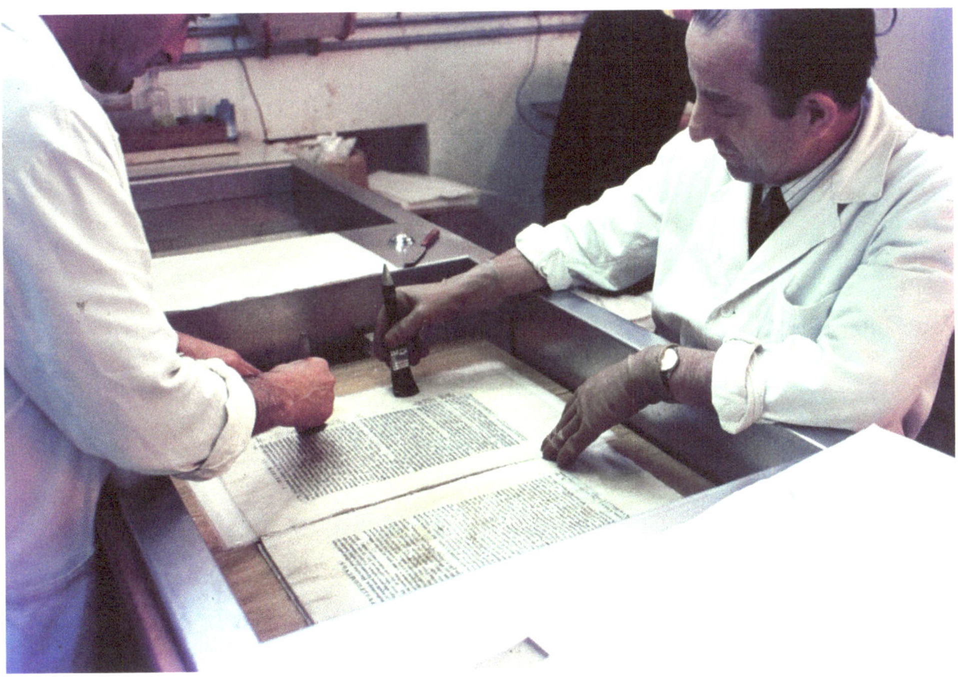

Figure 36 Washing department: leaves supported with interleaving and the mud streaks brushed away.

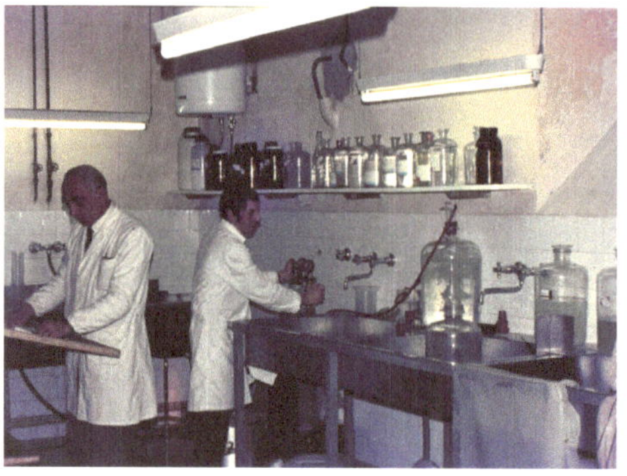

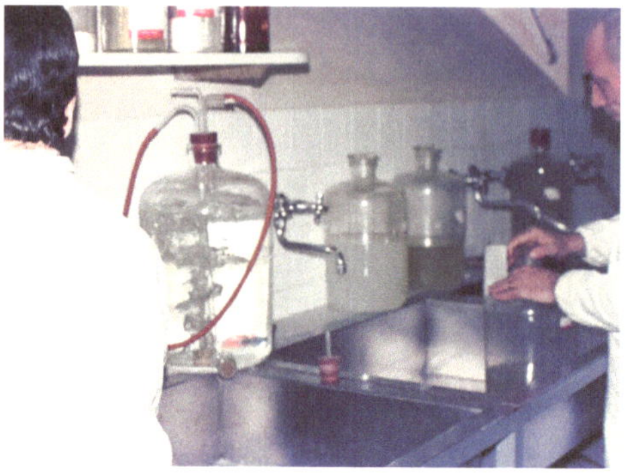

Figures 37 and 38 Italian staff preparing the Barrow buffering solution.

The bench designed by Peter Waters for making the heat-set tissue can be seen in Figure 39. At first, the resin was an internally plasticised PVA diluted with water, but later Joe Nkrumah introduced polymethacrylate dispersions which were tested by Margaret Hey as we had experienced blocking problems with the PVA. Figure 40 shows the stainless steel sink and hardwood support for the plate glass. The resin mixture was applied evenly onto the glass and a sheet of lens tissue bowed and lowered onto the resin and left to dry. Both sides of the plate glass were used. This casting method was based on Peter Water's experience of old photographic methods. The PVA introduced by Sandy Cockerell was superseded by Joe's glue and eventually became known as Library of Congress (LC) tissue following Margaret Hey's move to America. After yet another collation (Fig. 41), the completed text block was placed between pulp boards and wrapped in Topane-treated brown kraft paper and sent to the stacks.

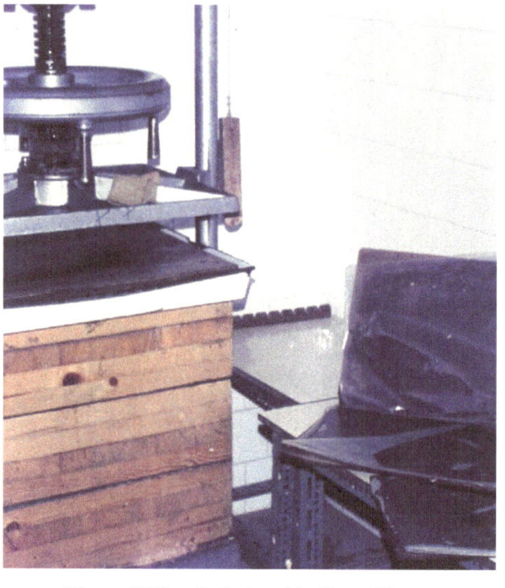

Figure 39 Bench designed by Peter Waters for making the heat-set tissue.

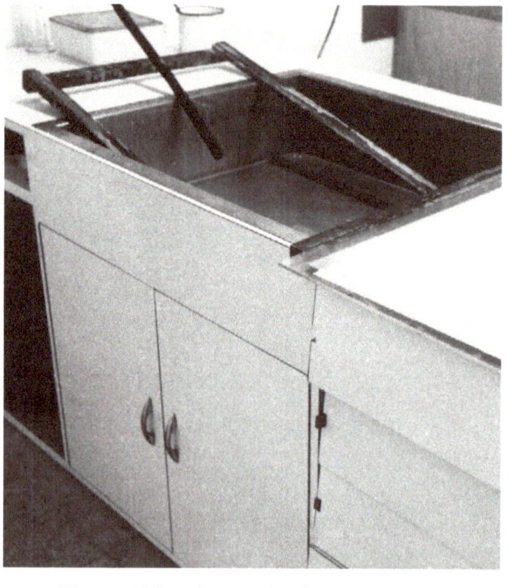

Figure 40 Stainless steel sink and hardwood support for the plate glass.

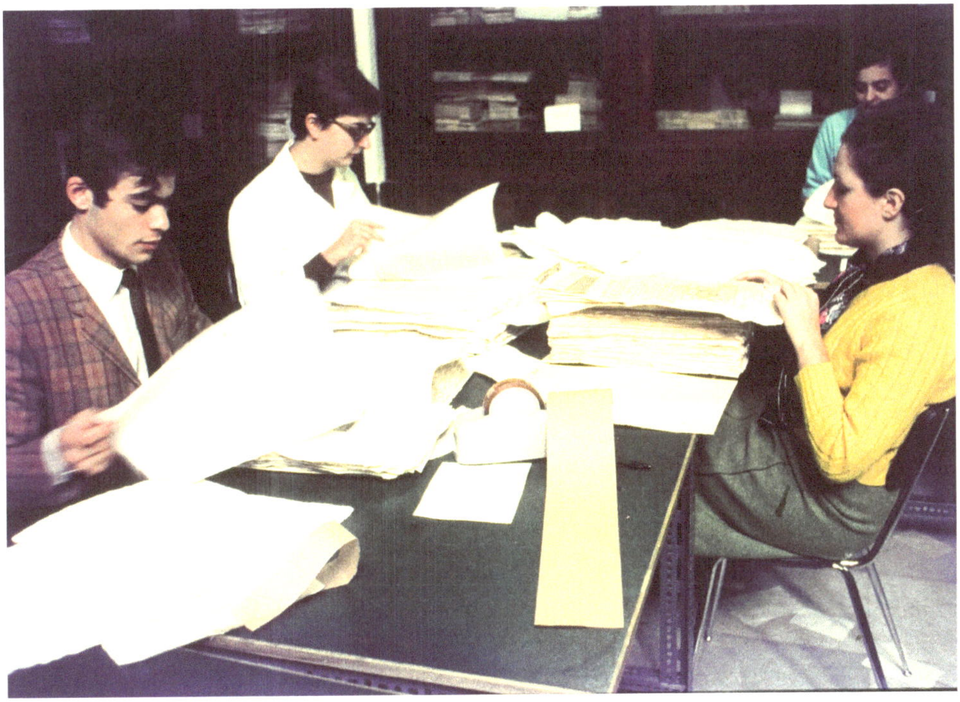

Figure 41 Collation of completed text block between pulp boards and wrapped in Topane-treated brown kraft paper.

Most of my time was spent at a workstation (Fig. 42). The figure also shows a later version of Peter and Sheila Waters' original specification card. The bibliographic details were prepared by Dr Crocetti, the binding components by Signor Daddi and Dr Bonanni. Christopher Clarkson's chart of limp vellum binding specifications can be seen in Figure 43 and his end-paper construction chart along with colour and material combinations for quarter bindings is illustrated in Figure 44. An example of a limp vellum binding by Christopher with his beautiful calligraphic lettering can be seen in Figures 45 and 46. A prepared skin of vellum made by Gentilli of Rome still in its stretching hoop is shown in Figure 47. A great deal of our effort was taken up with selection of suitable vellum (Fig. 48) and the treatment of vellum, alum-tawed and vegetable-tanned leather with the collaboration of Dr Don Reed (see Reed 1972 in the bibliography on p. 70).

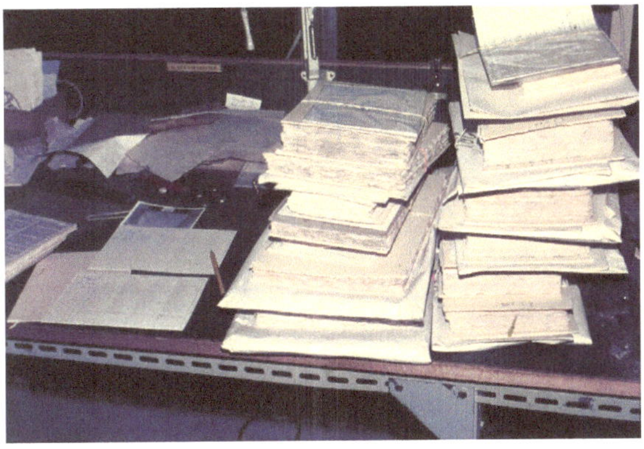

Figure 42 Anthony Cains' workstation with Peter and Sheila Waters' original specification card.

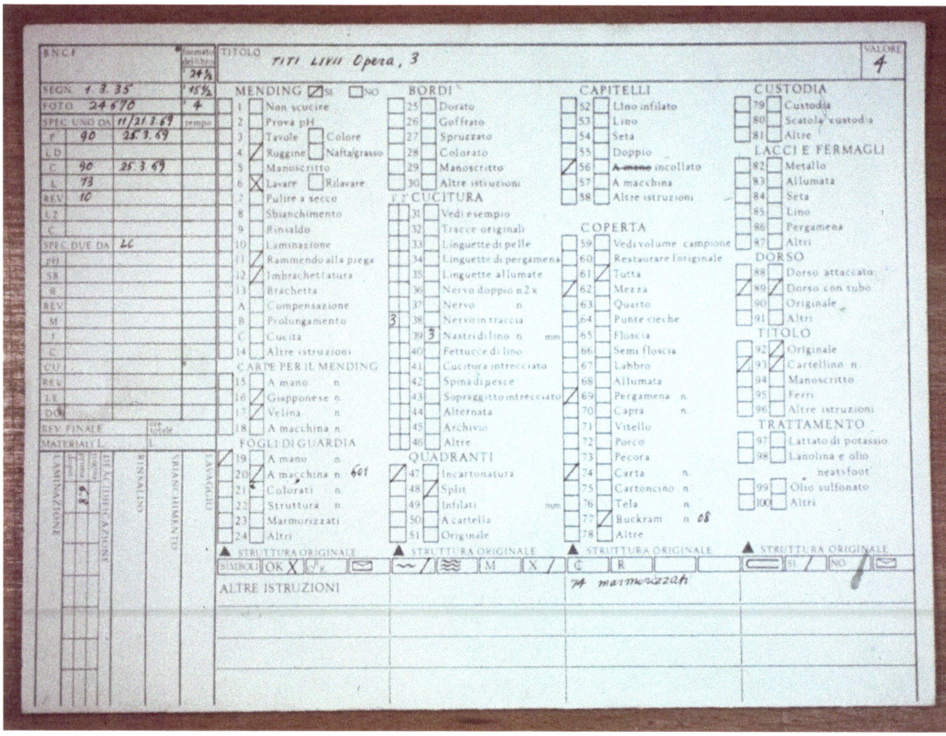

Figure 43 Christopher Clarkson's chart of limp vellum binding specifications and his end-paper construction chart.

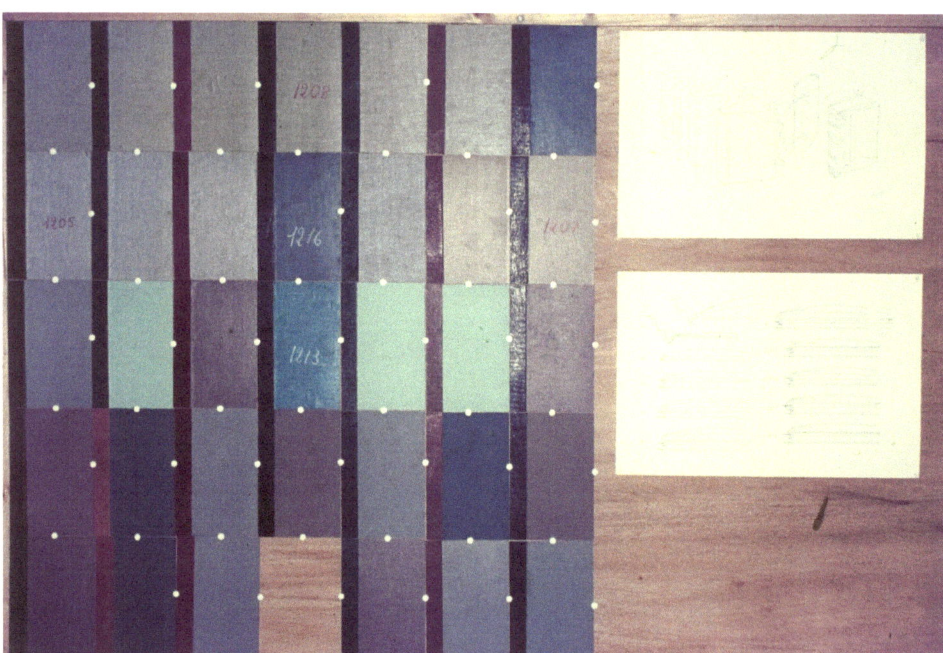

Figure 44 Colour and material combinations for quarter bindings.

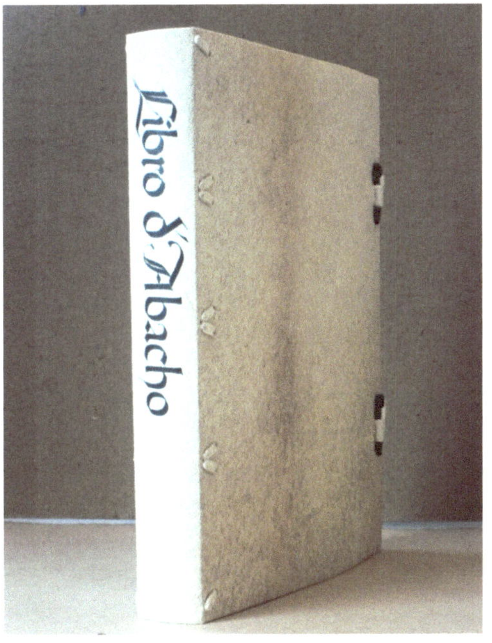

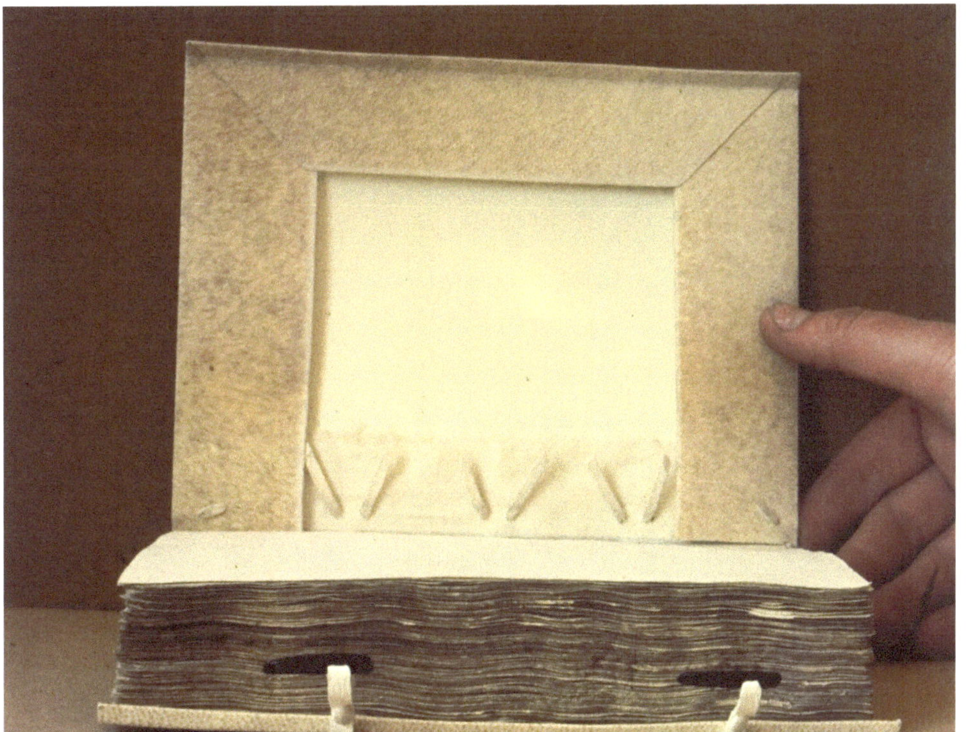

Figures 45 and 46 Limp vellum binding by Christopher Clarkson with his beautiful calligraphic lettering.

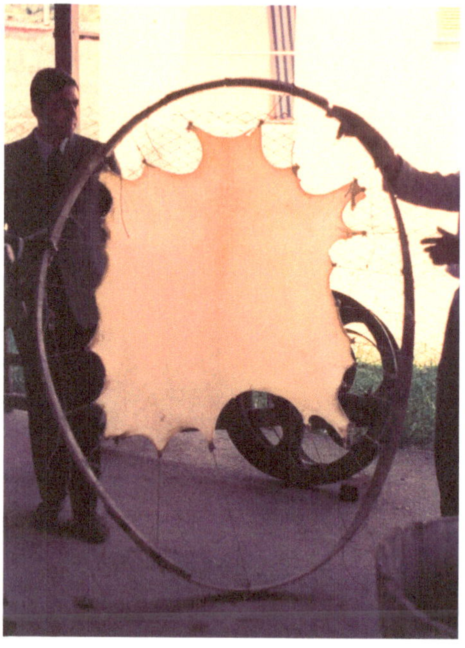

Figure 47 Prepared skin of vellum made by Gentilli of Rome, still on its stretching hoop.

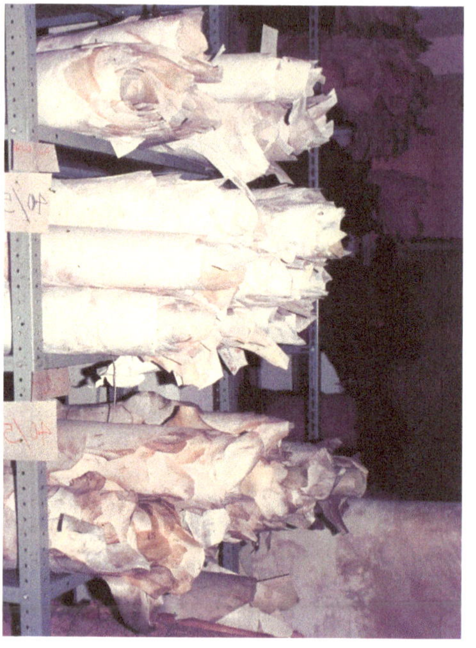

Figure 48 Parchment storage in the BNCF.

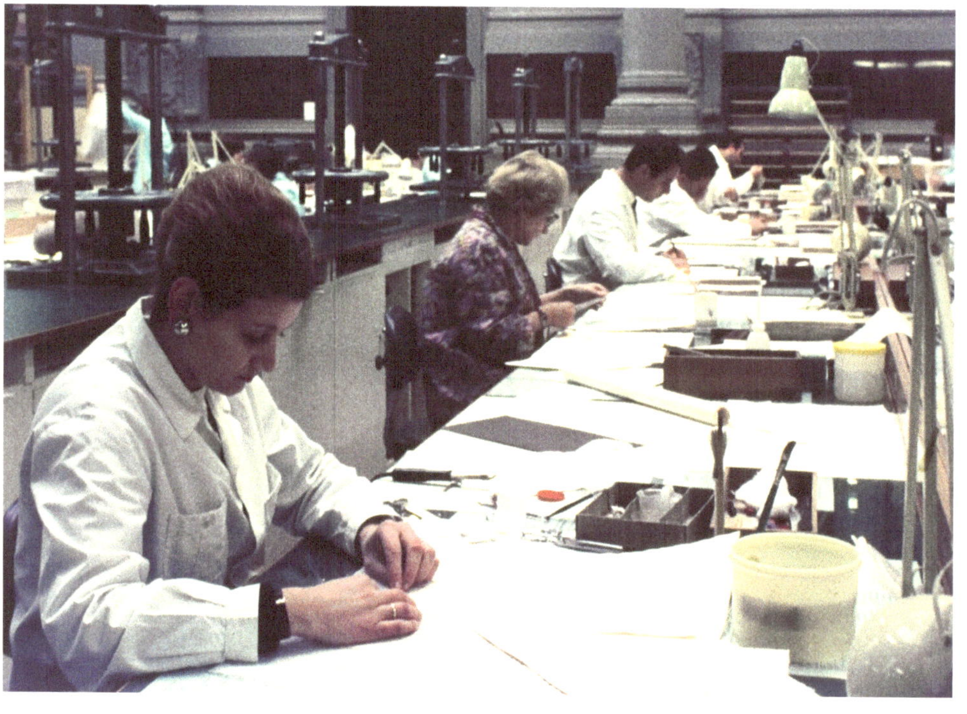

Figure 49 Mending department.

The mending department

A particularly striking feature of the design of the mending department (Fig. 49) is the way the percussion press base platen is set flush into the bench surface. The long fibre Japanese paper infilling technique (Figs 50–53), brought by Peter following a visit to Otto Waechter's workshop in Vienna, was combined with the heat-set tissue for edge repairs and weak areas of text. Figure 54 demonstrates the use of the temperature-controlled spatulas (designed by me and manufactured in St Albans, north of London). Experimentation was the culture of the system: Joe Nkrumah experimented with various cellulose derivatives, polyvinyl alcohol and starch; Dag-Ernst Petersen conducted some trials on the film-forming characteristics of various adhesives; and William Barrow carried out experiments on the combination of buffering and sizing of the text blocks for the BNCF. Finally, the book block was interleaved and pressed in sections to consolidate the repairs (Fig. 55).

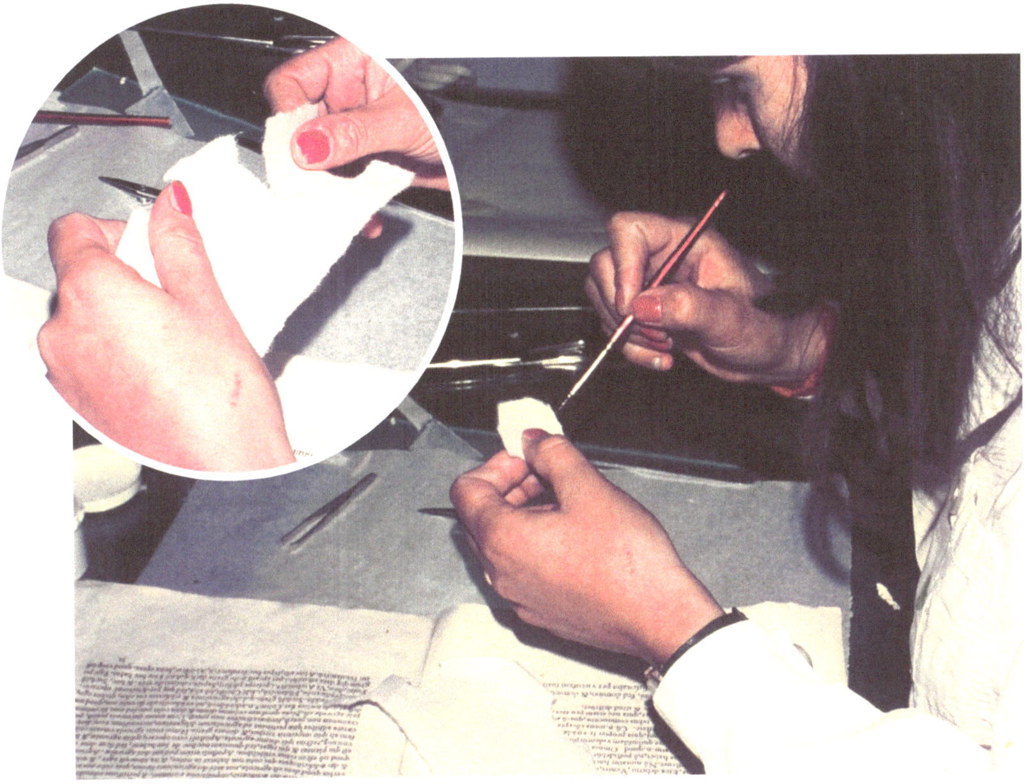

Figures 50 and 51 Long fibre Japanese paper infilling technique.

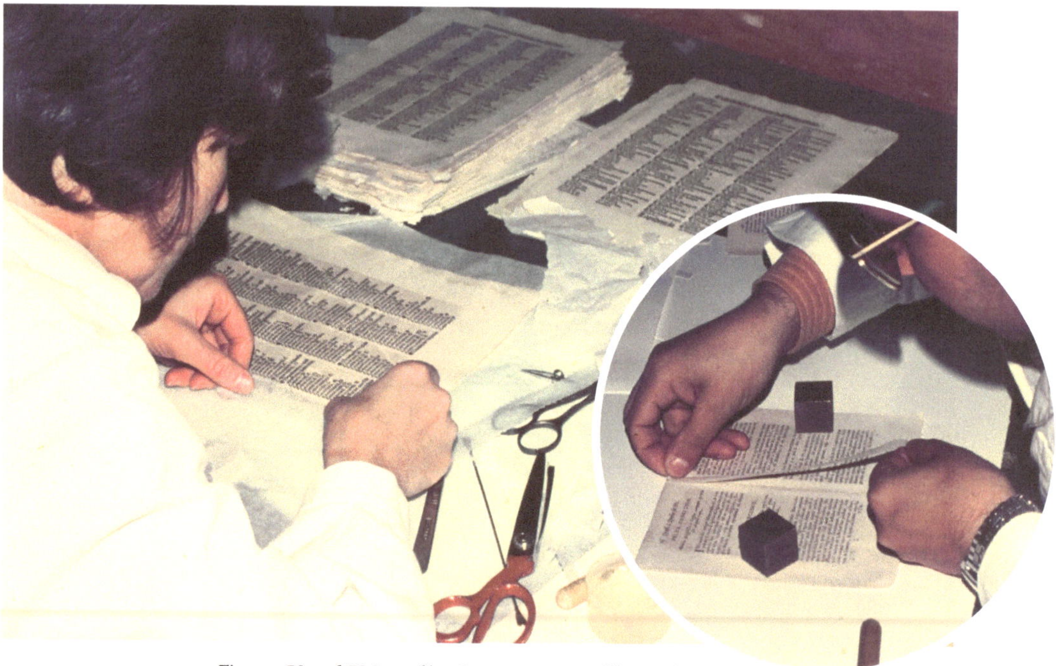

Figures 52 and 53 Long fibre Japanese paper infilling technique.

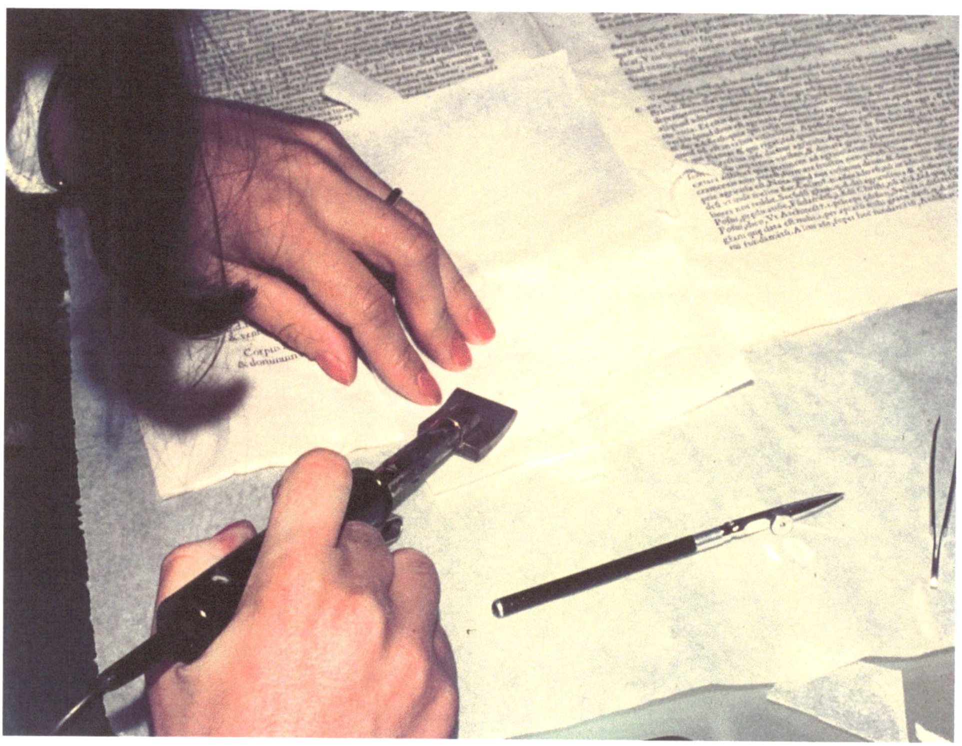

Figure 54 Temperature-controlled spatula used with heat-set tissue for edge tears and weak areas of text.

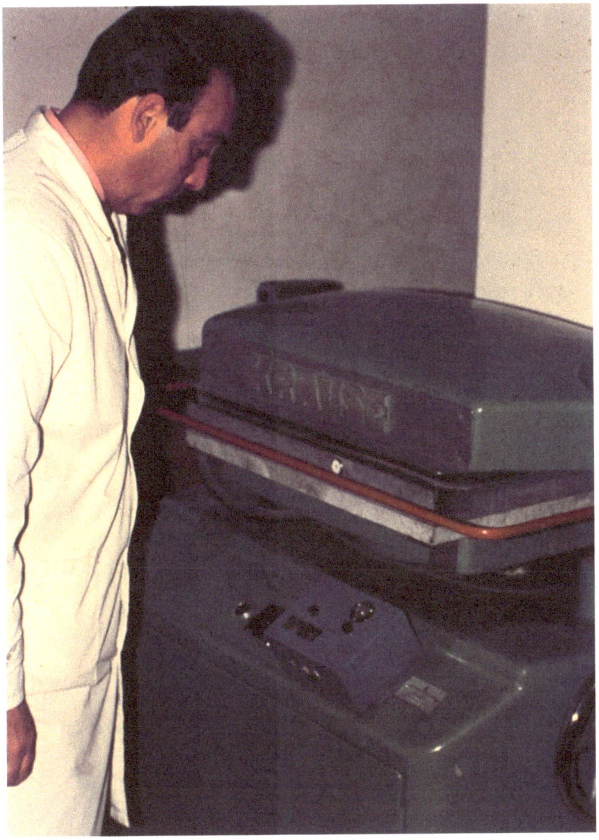

Figure 55 Book block interleaved and pressed in sections to consolidate repairs.

The sewing and binding department

The sewing and binding department is illustrated in Figure 56. The sewing was really good – a double flexible sewing on linen cords (Irish Linen netting twine by Barbour's, unobtainable today). No edge trimming, single or double cords (Figs 57 and 58) or alum-tawed thongs (Fig. 59) for limp vellum bindings and linen tapes for the less valuable material. These ancient methods are the foundation of good structure. Figure 60 is a picture of the bindery as it was set up initially before we were compelled to move to the basement area (Peter was not pleased!). In Figure 61, Christina Winburg from Sweden, Ada Rossi and Barbara Guiffrida can be seen working on the limp vellums. The original oak boards of a Hebrew text printed in Venice in the sixteenth century are illustrated in Figure 62. The text was sewn to the original mark-up and the boards laced on. Figures 63 and 64 illustrate the beautiful chevron end-band of the *Teuerdank* sewn by Dag-Ernst and Gabriele Petersen. They had a modern approach to covering, illustrated here by the formation of the tawed cover at the back edge (*Epoca* 1971). The core slips of the end band were laced in order to avoid slitting the turn-in along the hinging joint, creating weakness. Dag-Ernst made the cut so that it fell onto the tight back spine, hence the part-covered cap form. For paring the leather, we sometimes used a Fortuna SAS machine (a common piece of equipment in many large workshops) (Fig. 65) but more often this was done by hand using various traditional knife types or with a Stanley metal spoke shave (Fig. 66). Figure 67 shows a sixteenth/seventeenth-century red goatskin leather cover being infilled with patches of modern goatskin. The joint area was intact so the book block had to be built exactly to fill it, which was tricky. The cover has been hydrated, treated with sulphonated oil (Reed 1972) and manipulated, i.e. stretched (or compressed) to fit the dimensions of the sewn and boarded book. The forwarded book was then checked for its alignment with the cover (Fig. 68). Although it is sewn on raised cords it has been fitted with a type of hollow lining, a flat tube of paper rather than being glued directly to the spine folds. The next figure (69) is of two large volumes, forwarded and ready for covering in a new vegetable-tanned leather, followed by another folio (Fig. 70), which has a new, red Nigerian goatskin spine, matching the original leather, and finally a full tawed goatskin binding in the press having had its spine tied with cord (Fig. 71), as seen earlier with Dag-Ernst's work. This ensures the leather is firmly fixed to the spine, at the same time giving a 'period' look to the binding. The white tawed goatskin, also by Gentilli of Rome, is a really beautiful and durable leather. The following sequence (Figs 72–75) illustrates the completed work. The figures show limp vellums, full tawed and tanned leather covers, and quarter leather representative of four of the five 'period' specifications:

- Value 1 = 1500–1560
- Value 2 = 1500–1770
- Value 3 = 1770–1820
- Value 4 = 1820–1840
- Value 5 = post-1840

Figure 76 shows the function of a throw-out guard on an illustrated text. These are finished bindings back in the stacks; some are just re-backs but many have retained their original spine covers. The grand format volumes shown in Figures 77 and 78 in green gold-tooled goatskin were bound by the Danes and were some of the earliest to be repaired in the system.

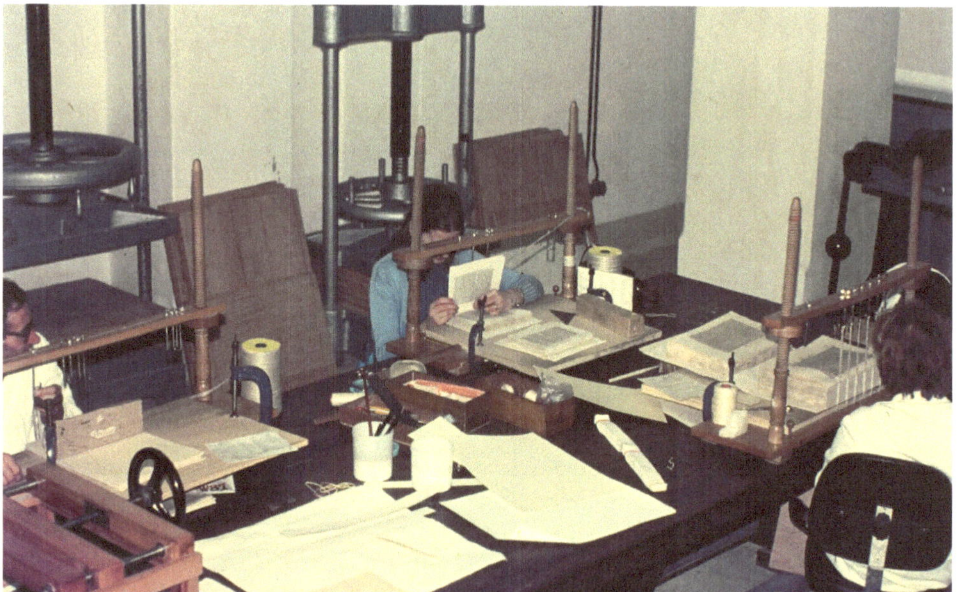

Figure 56 Sewing and binding department.

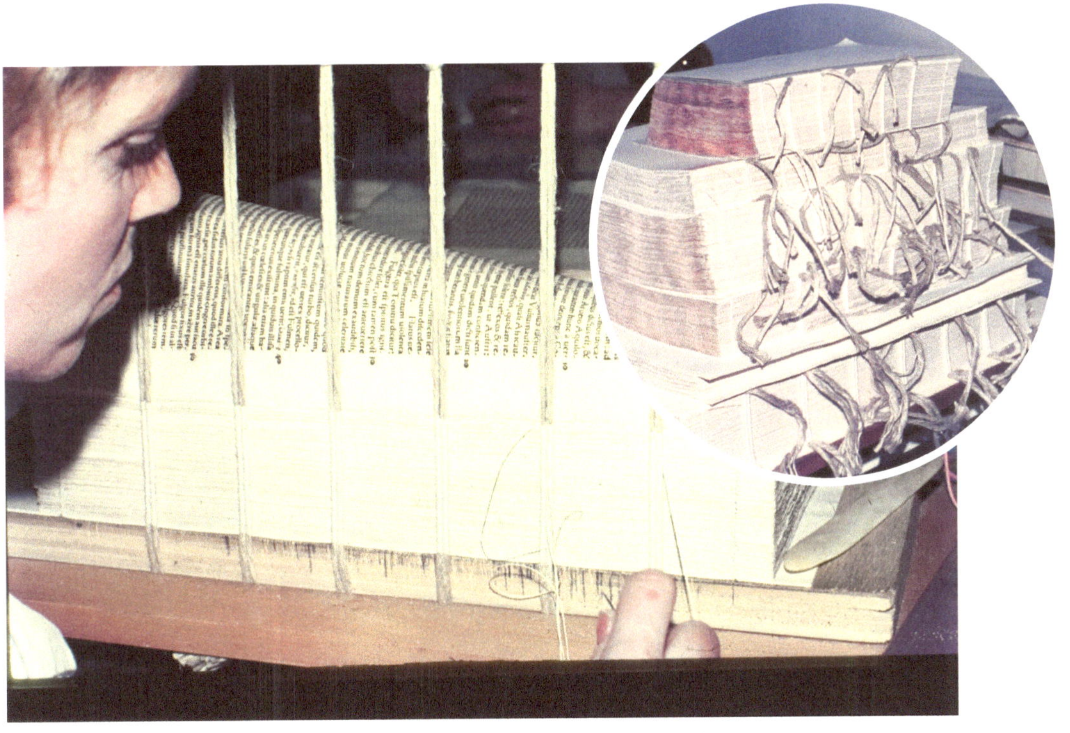

Figure 57 and 58 Books sewn on single and double cords.

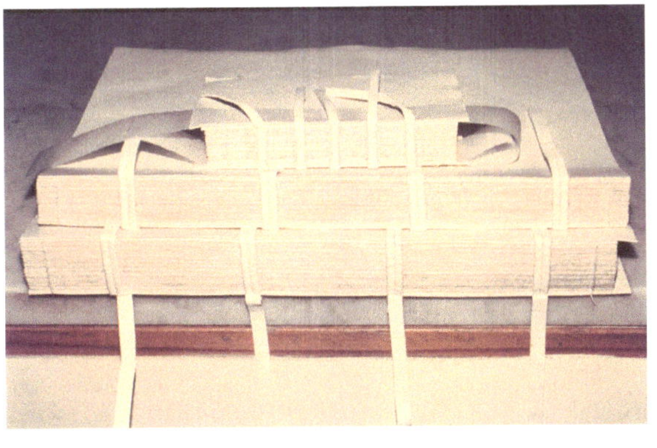

Figure 59 Sewing on alum-tawed bands.

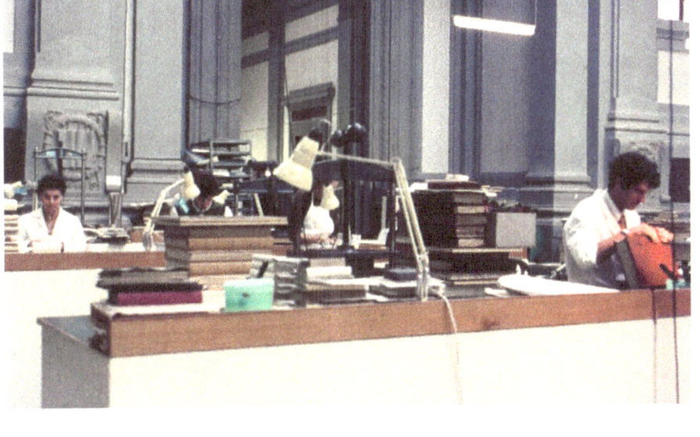

Figure 60 The bindery.

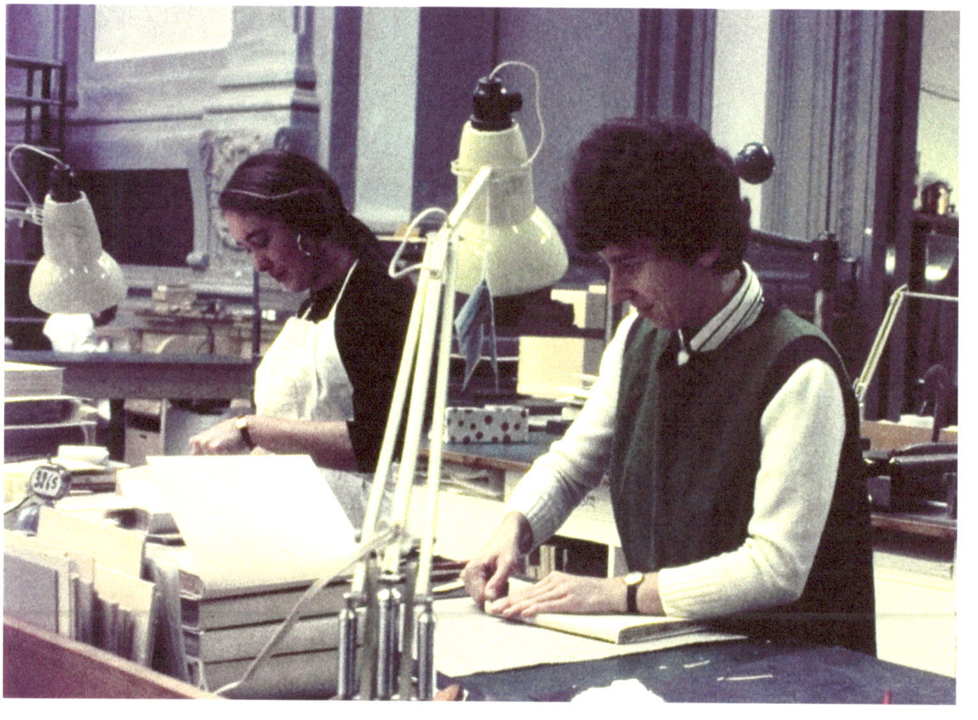

Figure 61 Christina Winburg (Sweden), Ada Rossi and Barbara Guiffrida.

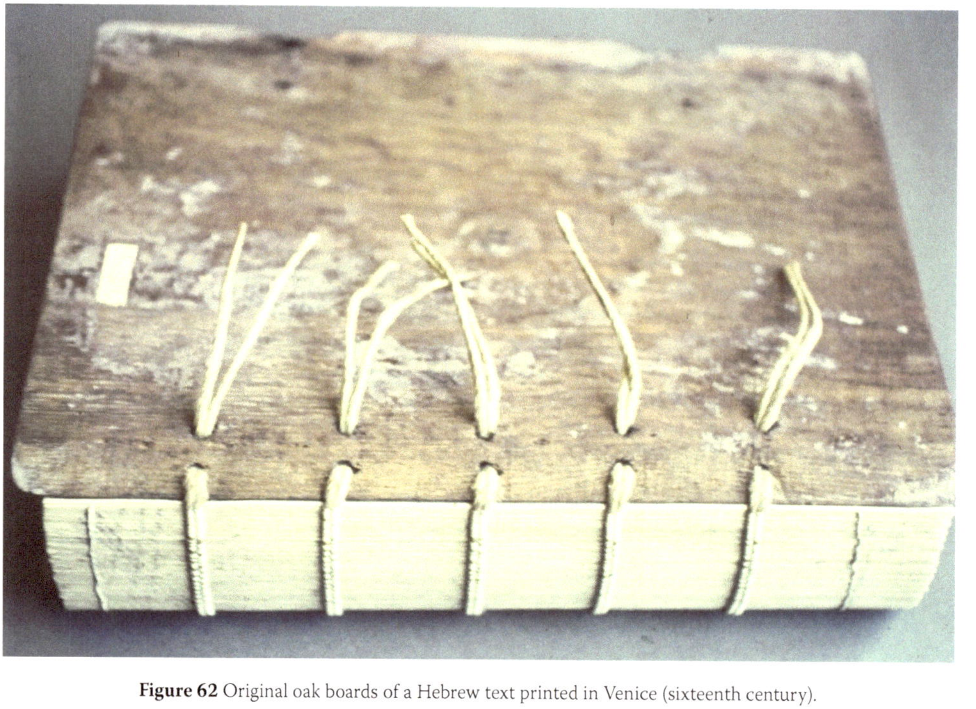

Figure 62 Original oak boards of a Hebrew text printed in Venice (sixteenth century).

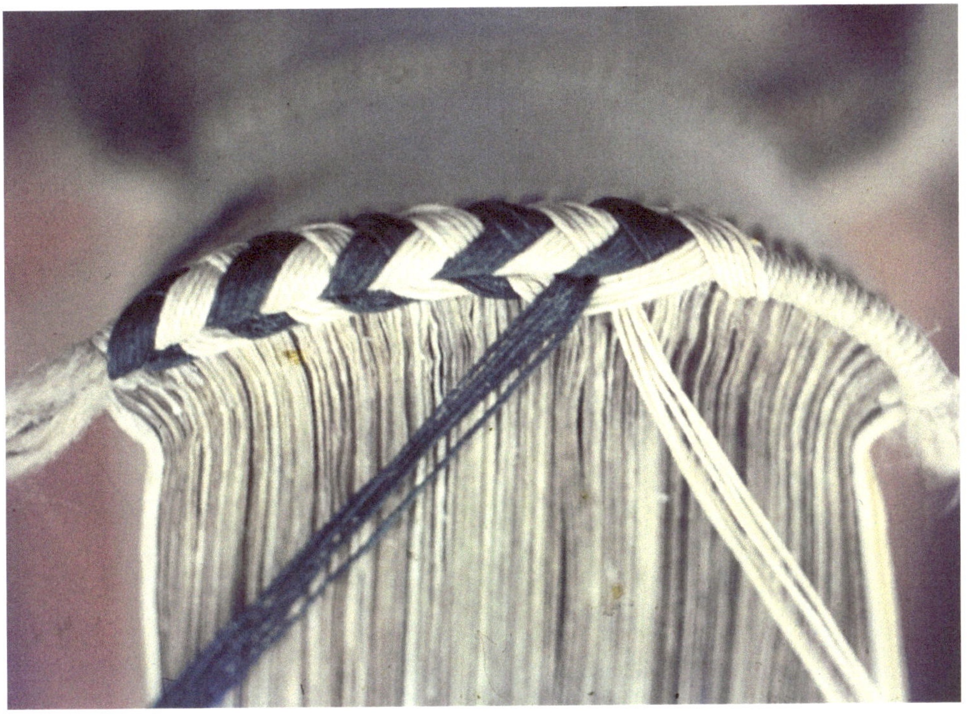

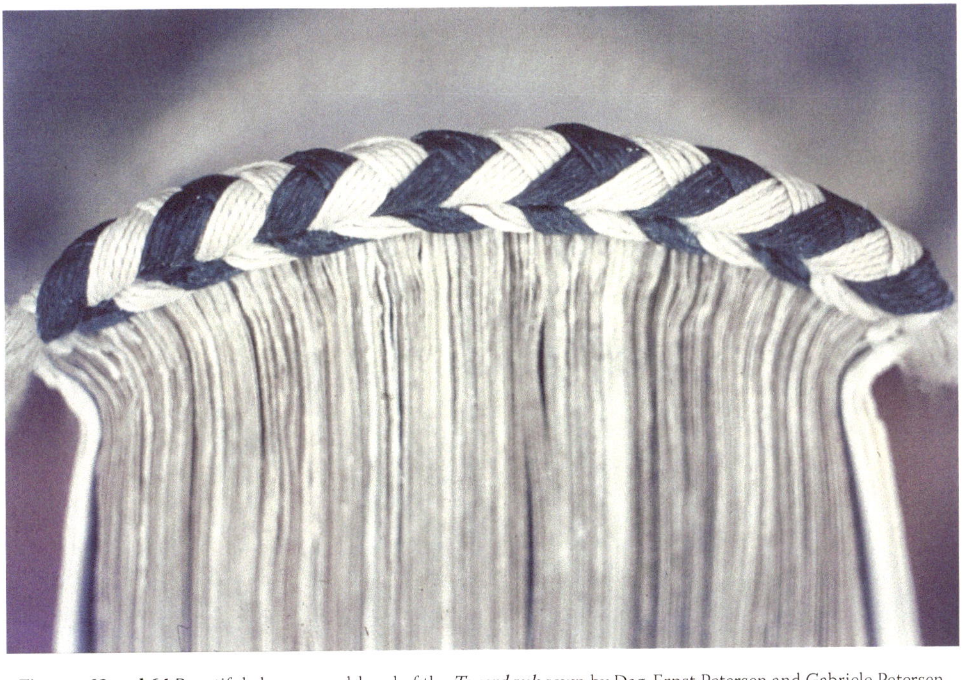

Figures 63 and 64 Beautiful chevron end-band of the *Teuerdank* sewn by Dag-Ernst Petersen and Gabriele Petersen.

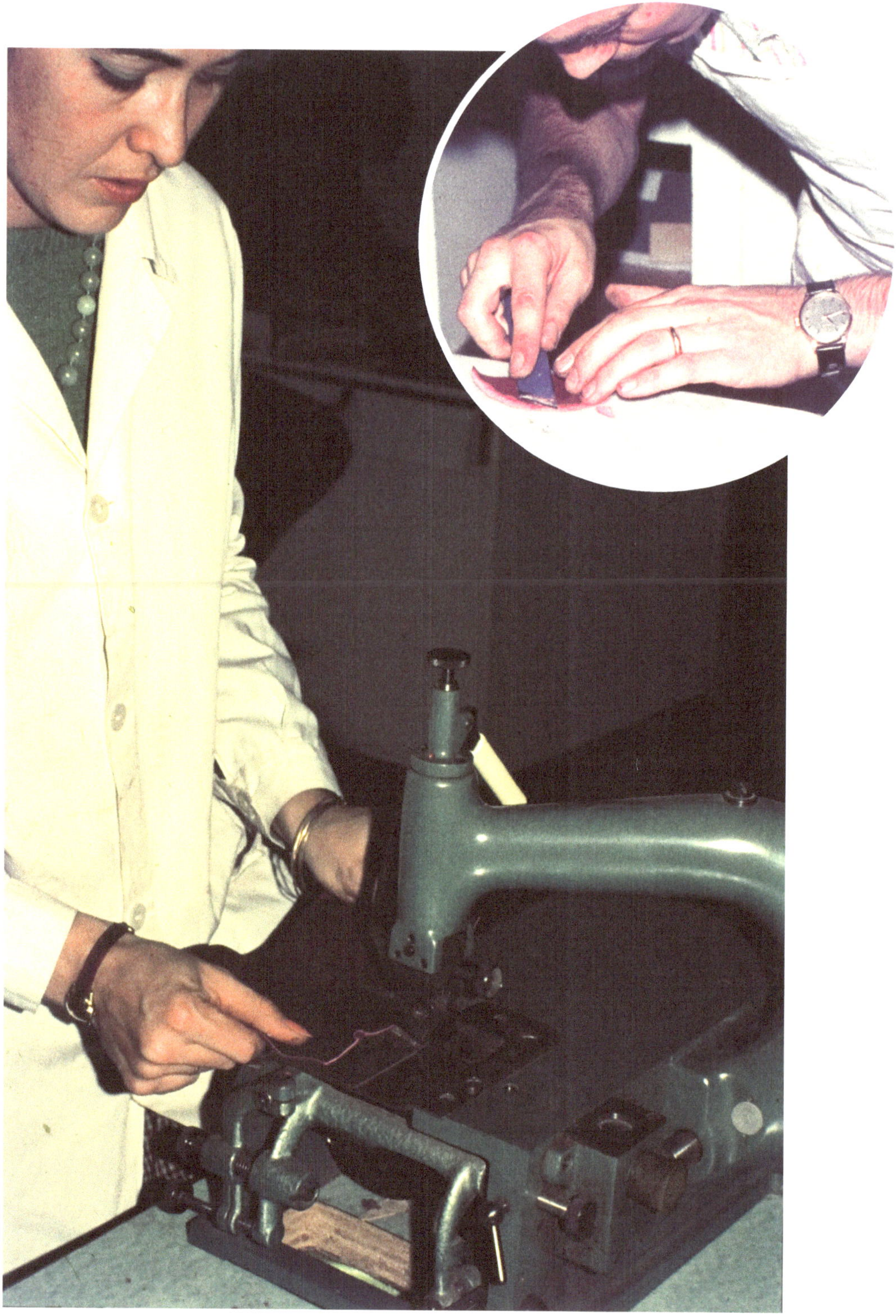

Figure 65 Fortuna SAS machine used for paring leather.
Figure 66 (top right) Stanley metal spoke shave.

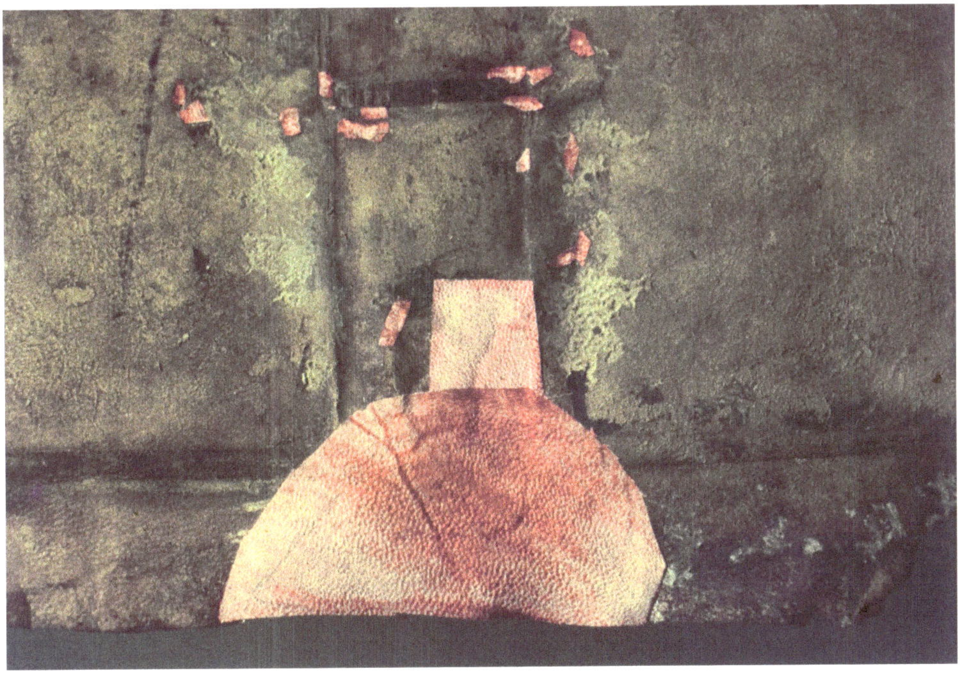

Figure 67 Sixteenth/seventeenth-century red goatskin leather cover being infilled with patches of modern goatskin.

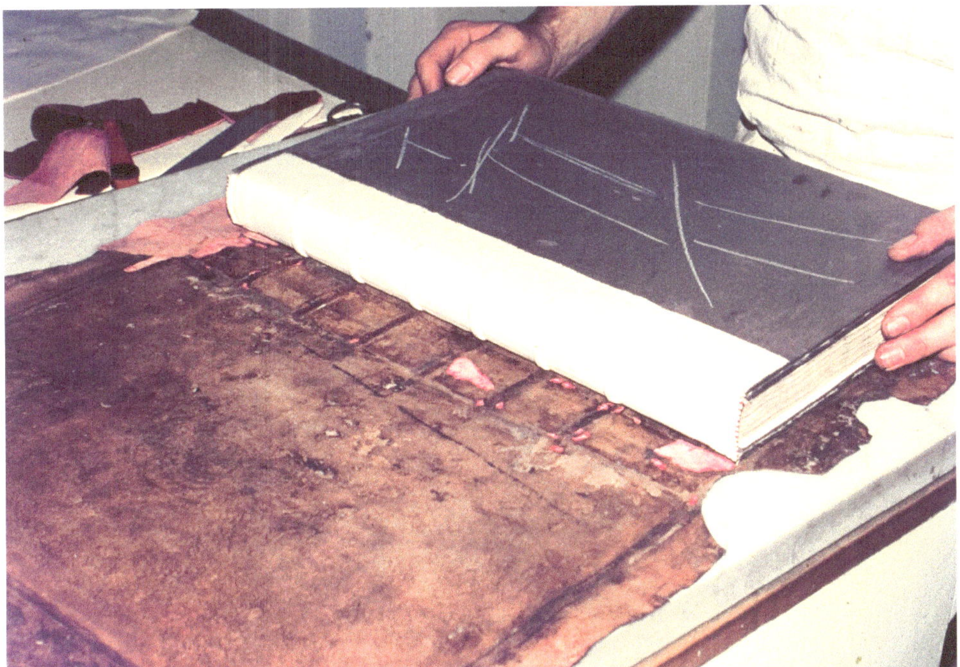

Figure 68 The forwarded book being checked for its alignment with the cover (*Sunday Times Magazine* 11 December 1966).

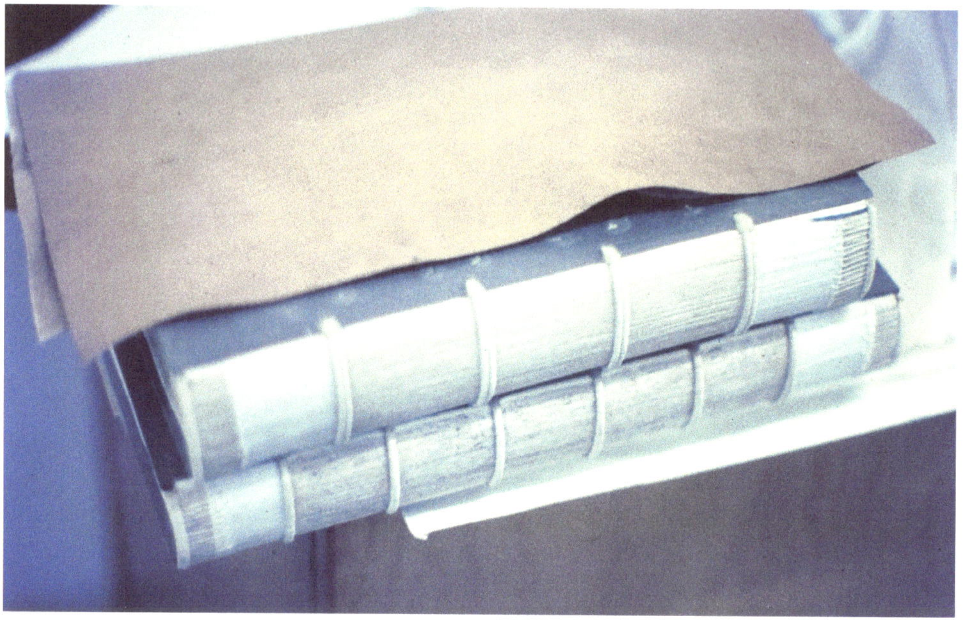

Figure 69 Two large volumes forwarded and ready for covering in new vegetable-tanned leather.

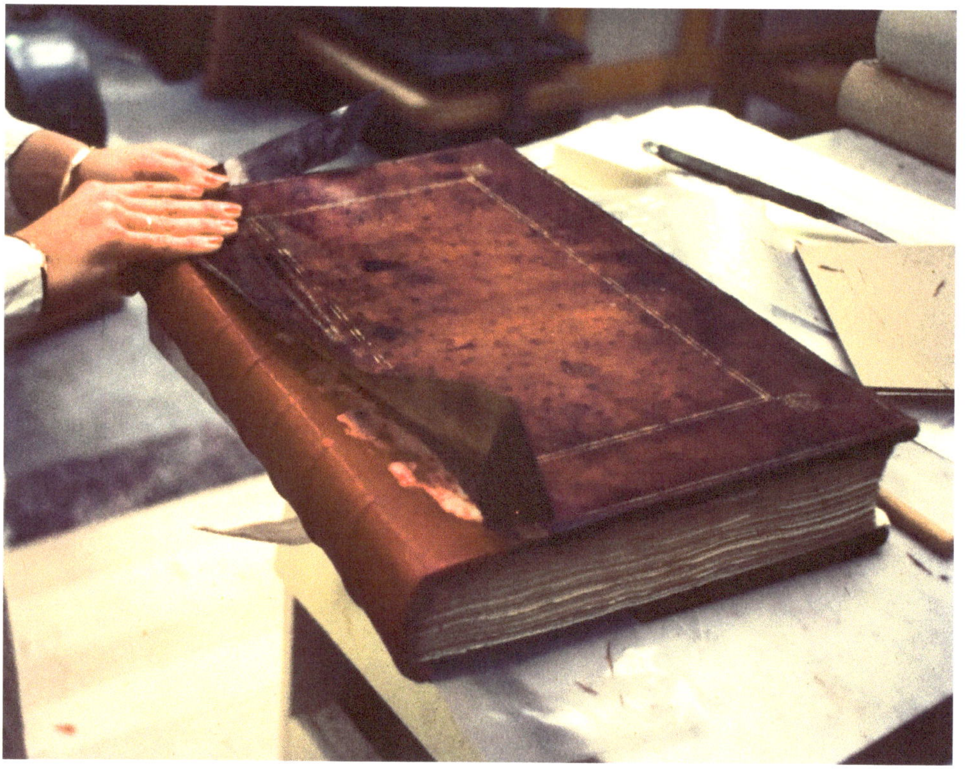

Figure 70 A folio with a new red Nigerian goatskin spine, matching the original leather.

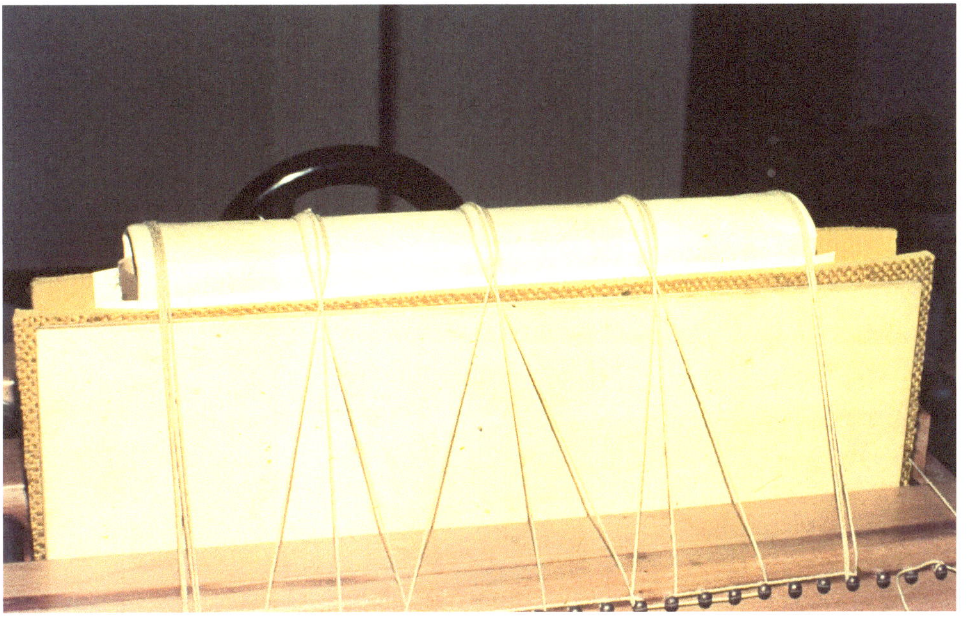

Figure 71 A full tawed goatskin binding in the press having had its spine tied with cord.

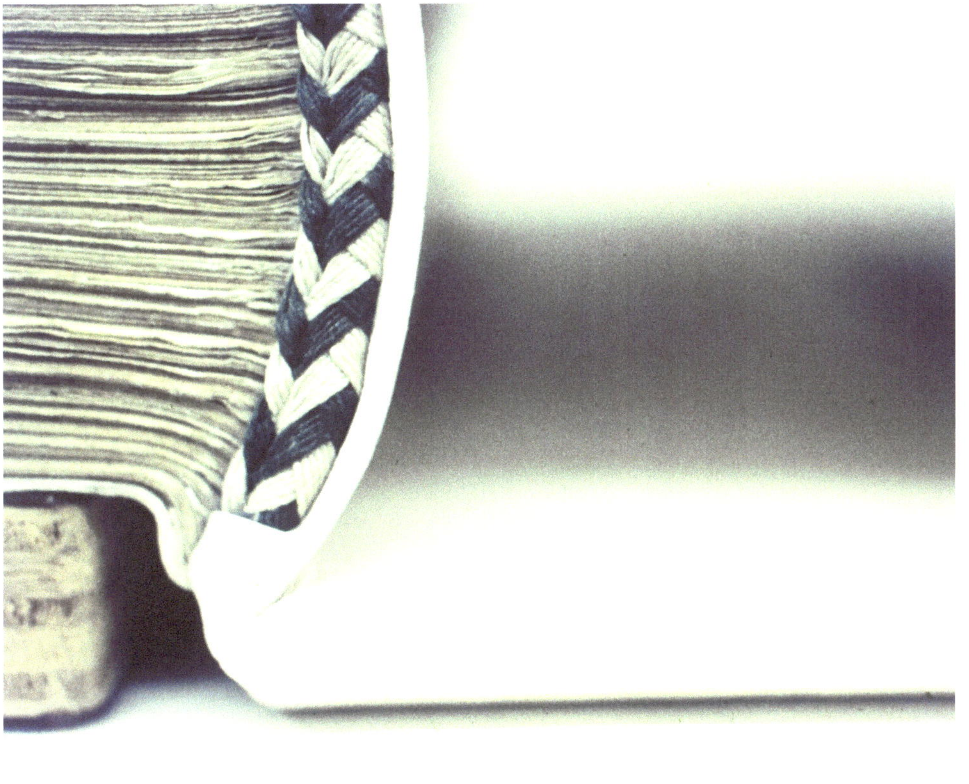

Figure 72 Completed work.

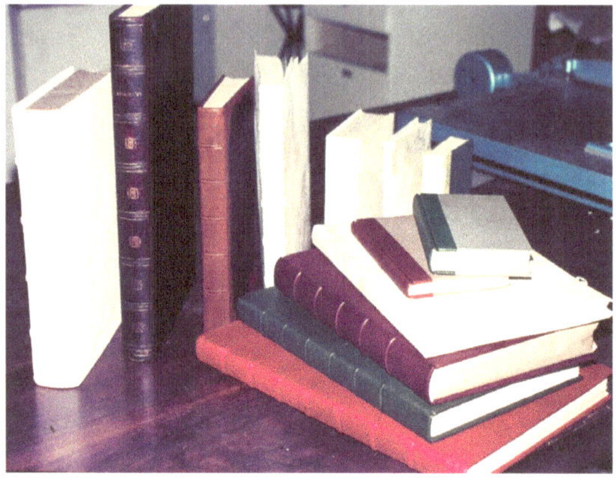

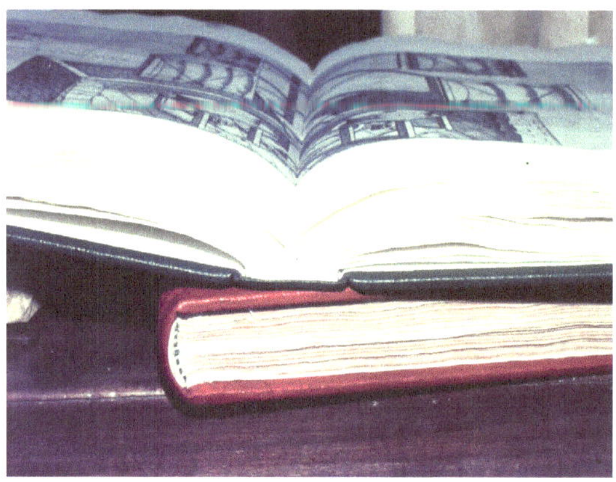

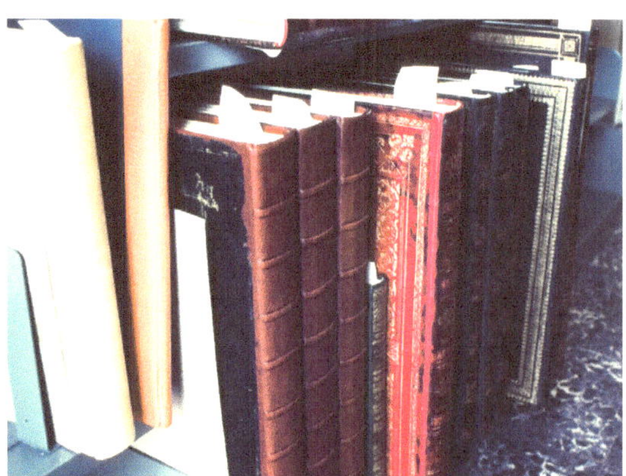

Figures 73, 74 and 75 Completed work.

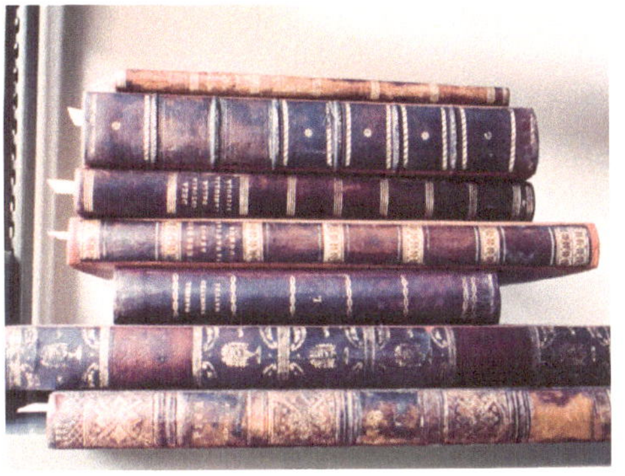

Figure 76 Finished bindings back in the stacks.

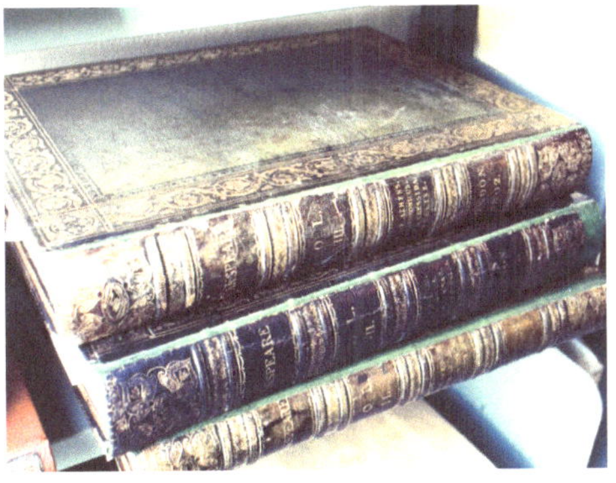

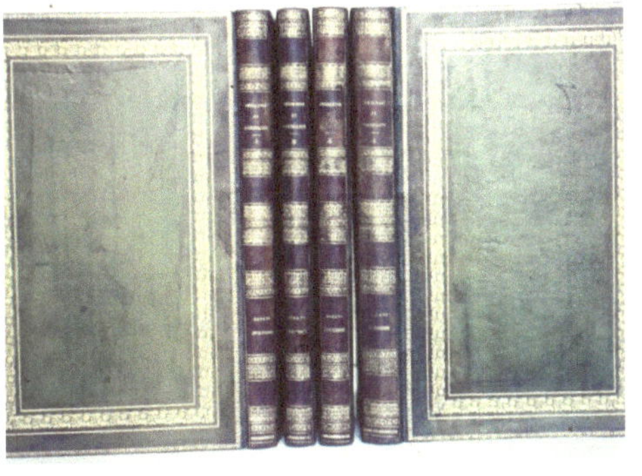

Figures 77 and 78 Grand format volumes in green gold-tooled goatskin bound by the Danes.

Restauro di Stampe

The Restauro di Stampe was set up by William Boustead with funds donated by the government of New South Wales. Joe Nkrumah can be seen in Figure 79 in consultation with a conservator from Prague, whose name I sadly do not recall. Franca Manera and Claudio Montelatici (Fig. 80) are seen probably spraying stained paper with ethereal hydrogen peroxide. Figure 81 illustrates a trial sheet of sample papers prepared by Margaret Hey in order to find the optimum pH for bleaching with freshly prepared sodium hypochlorite, followed by a sequence illustrating the process to remove the black tannin stains (Figs 82–86). After treatment with anti-chlor (sodium thiosulphate), the object received a lengthy wash in running water. After drying, it would be sized with methyl cellulose and repaired, ready for binding.

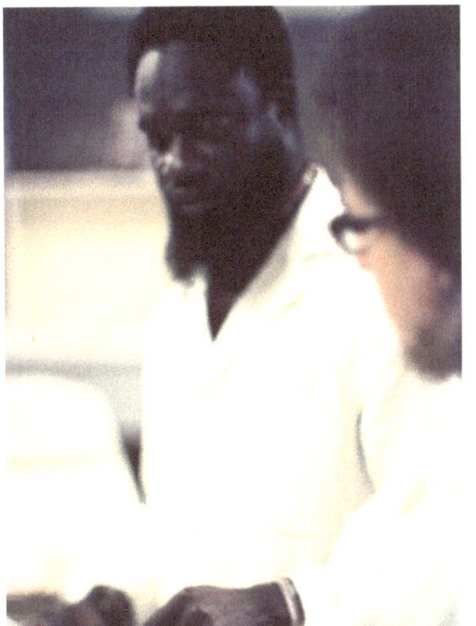

Figure 79 Joe Nkrumah in consultation with a conservator from Prague.

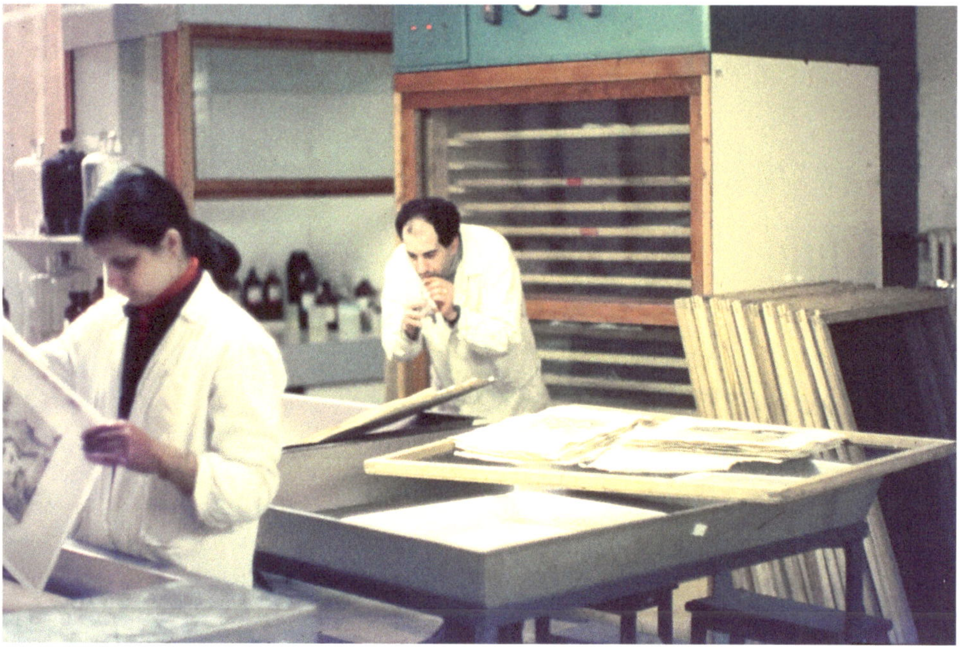

Figure 80 Franca Manera and Claudio Montelatici, who is probably spraying stained paper with ethereal hydrogen peroxide.

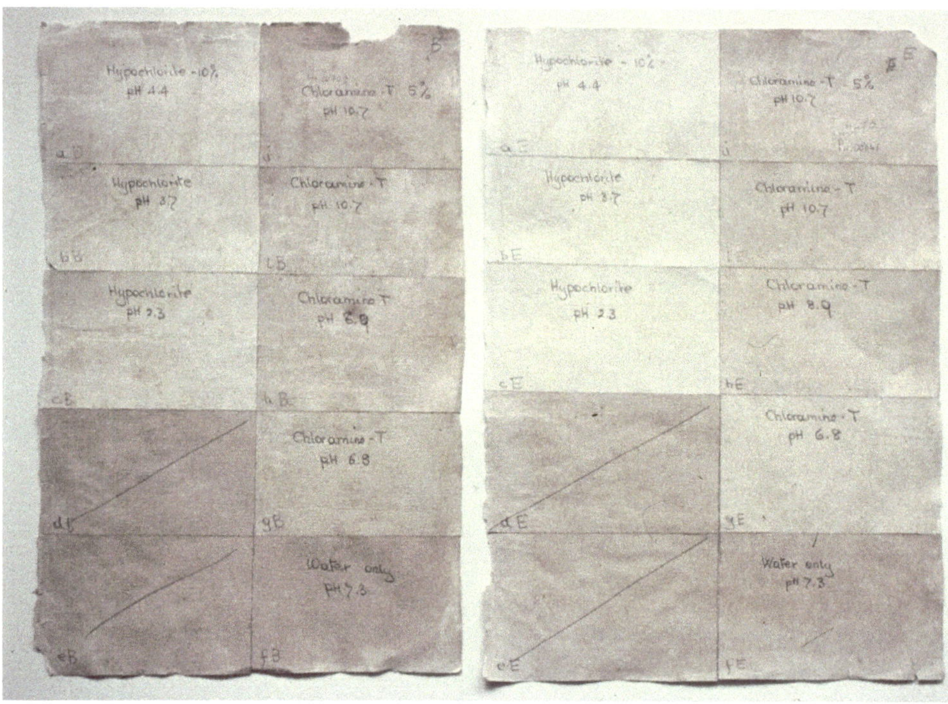

Figure 81 Trial sheet of sample papers prepared by Margaret Hey in order to find the optimum pH for bleaching with freshly prepared sodium hyperchlorite.

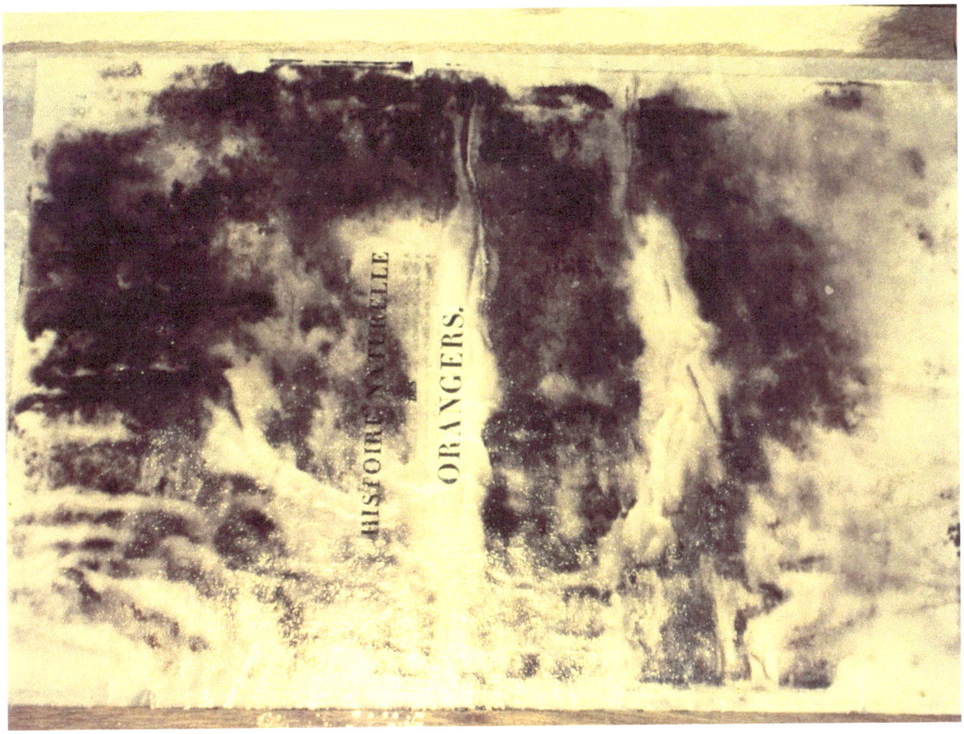

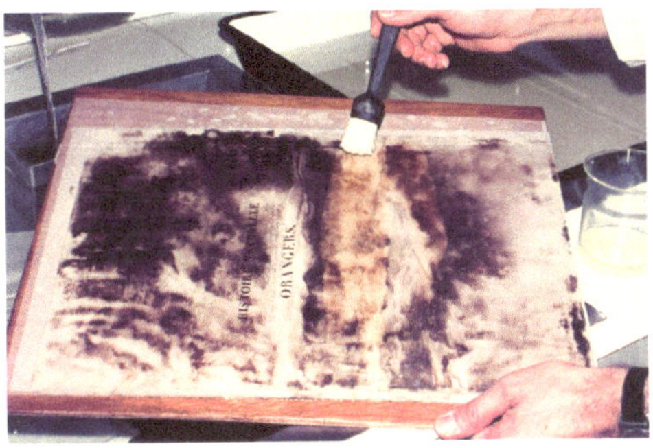

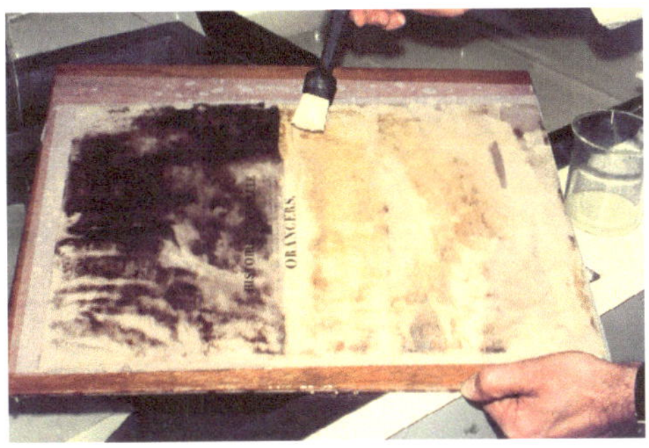

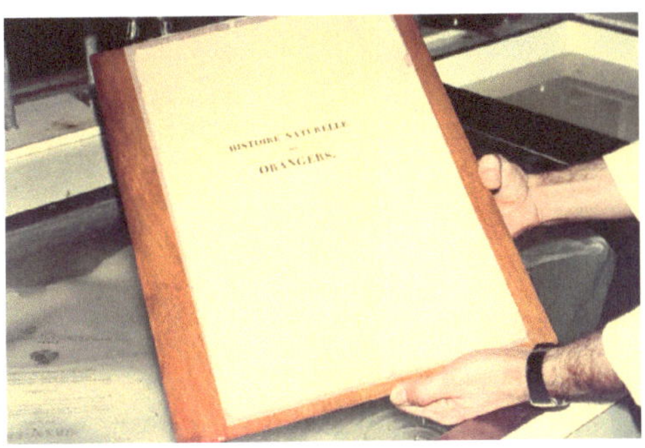

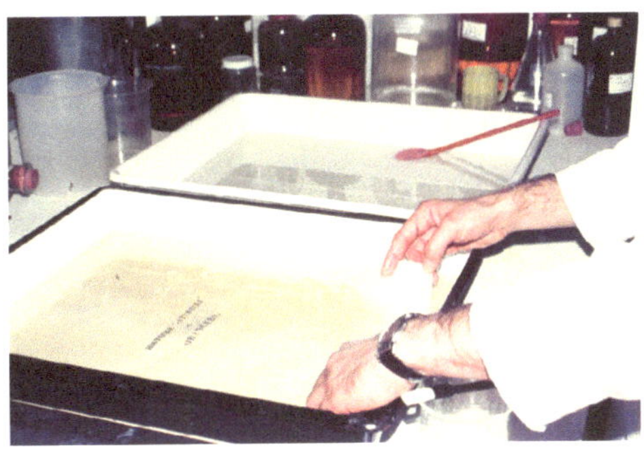

Figures 82, 83, 84, 85 and 86 Sequence illustrating the process to remove the back tannin stains.

The mechanised bindery (value 5)

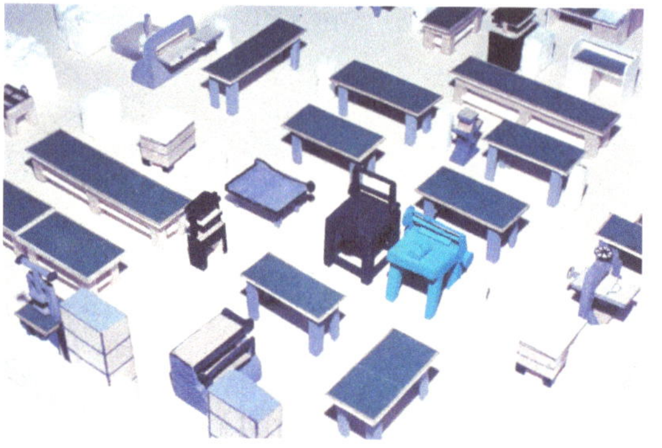

Figure 87 Model prepared by Mammo Bogale of every item of machinery and workbenches to determine their best location.

A small industrial unit was established by Herr W. Bleicher of Munich, which was enlarged for rebinding modern books in full cotton buckram. A model was prepared by Mammo Bogale (Fig. 87) of every item of machinery and workbenches to determine their best location. The following sequence of figures illustrates the text repair; hand sewing on tapes (Figs 88 and 89); the Martini book-sewing machine (Fig. 90); a batch of machine-sewn books (Fig. 91); the Ledueil roller backing machine (Fig. 92) and a group of treated books (Fig. 93); a powered board shear (Fig. 94); the pre-cut materials for spine linings (Fig. 95); books boarded ready for covering (Fig. 96); covering and finally lettering in gold foil by Roberto Bartolini (Figs 97–100). The final two illustrations are of Carla Bonani and Luigi Crocetti (Fig. 101) to whom I reported during a diplomatic visit to Rome. This is the only photograph I have of Professor Cassamassima (Fig. 102), our great leader. We can be seen in Milan looking at yet another high-tech solution for washing mud-impregnated paper which didn't work – the sample simply disintegrated!

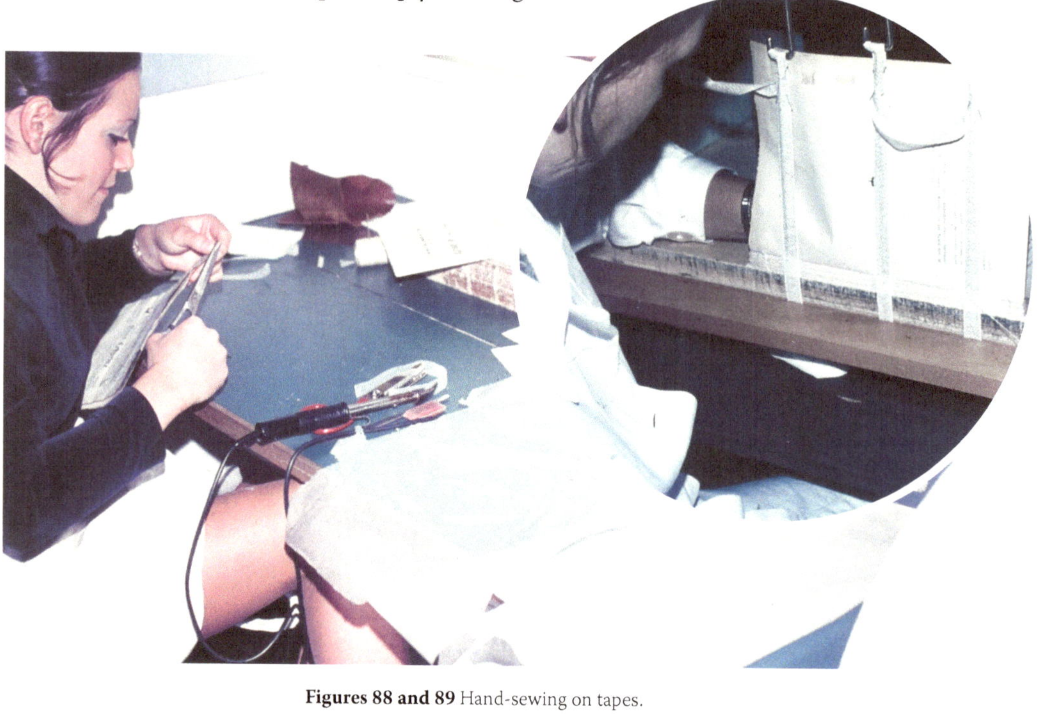

Figures 88 and 89 Hand-sewing on tapes.

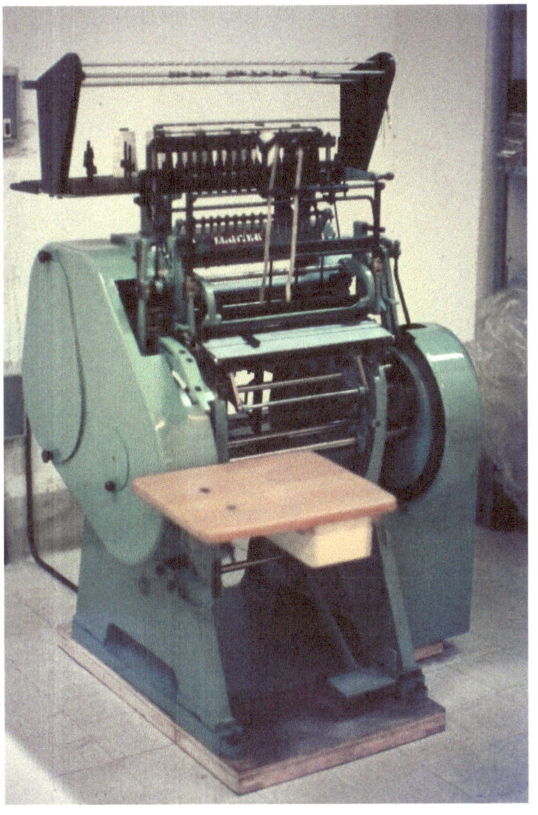

Figure 90 Martini book-sewing machine.

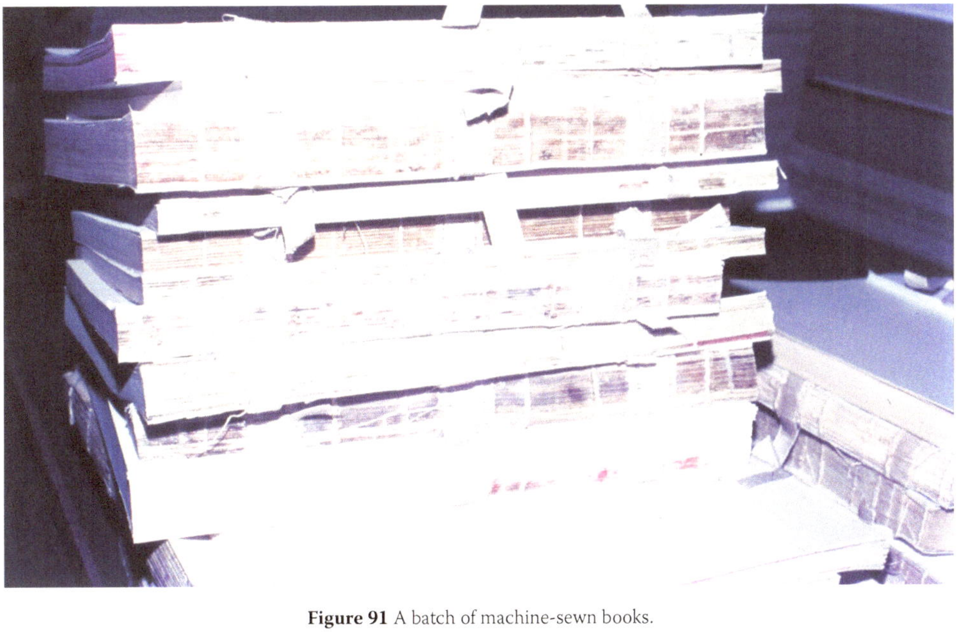

Figure 91 A batch of machine-sewn books.

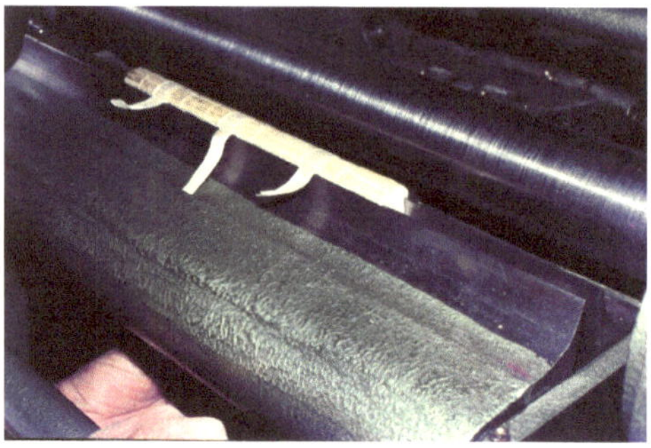

Figure 92 Ledueil roller backing machine.

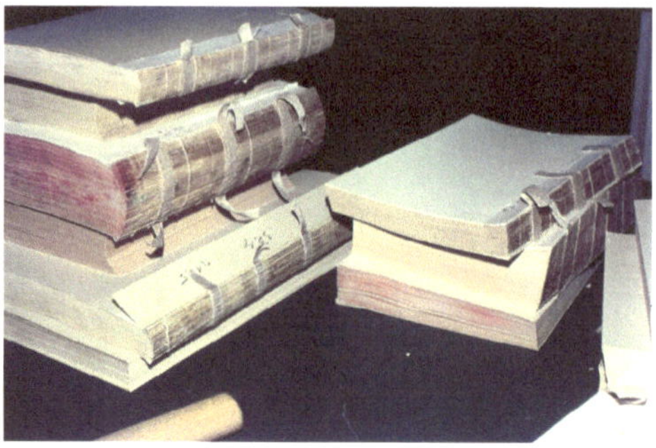

Figure 93 A group of treated books.

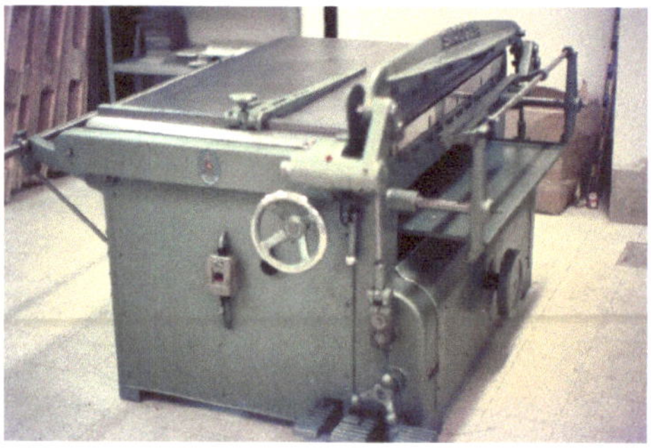

Figure 94 A powered board shear.

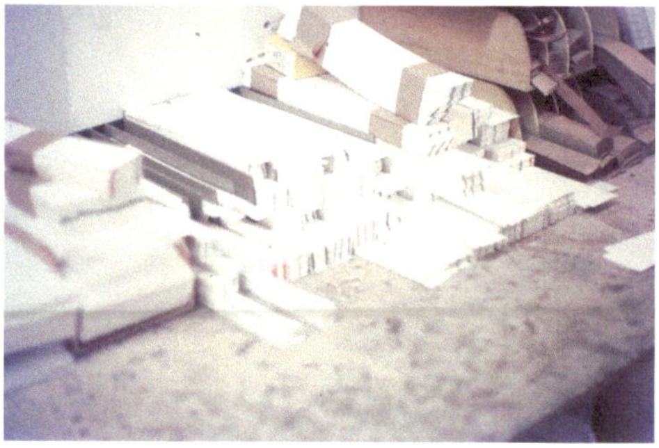

Figure 95 Pre-cut materials for spine linings.

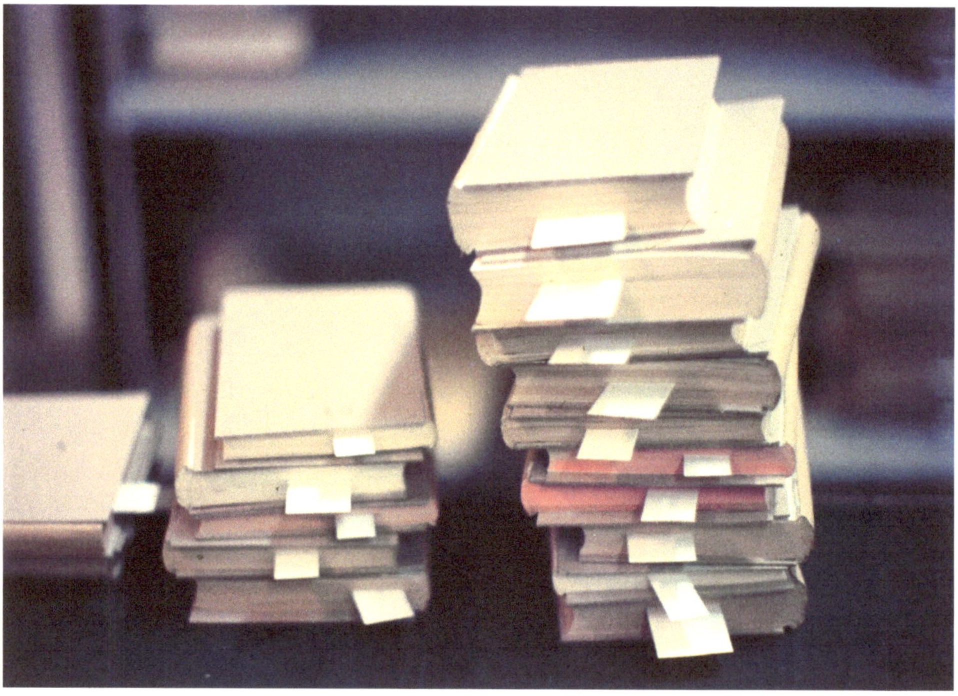

Figure 96 Books boarded ready for covering.

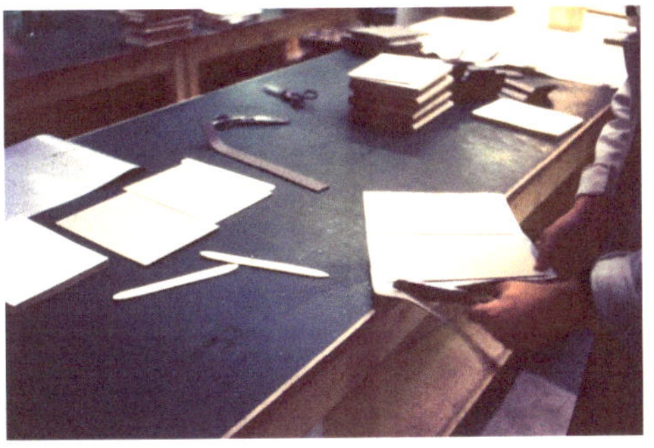

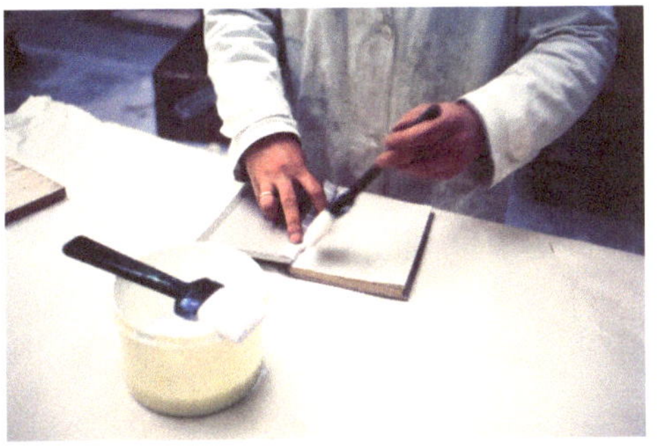

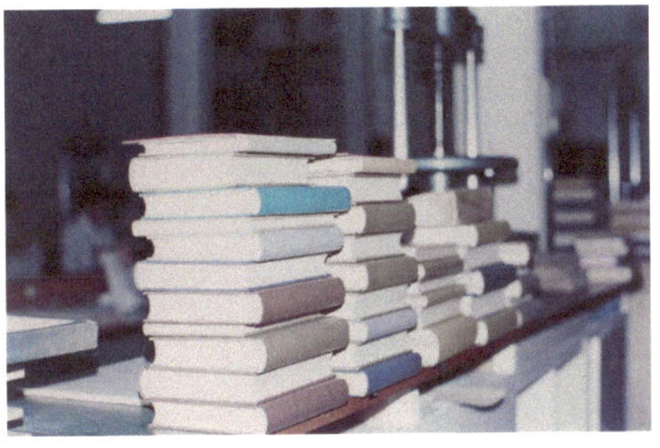

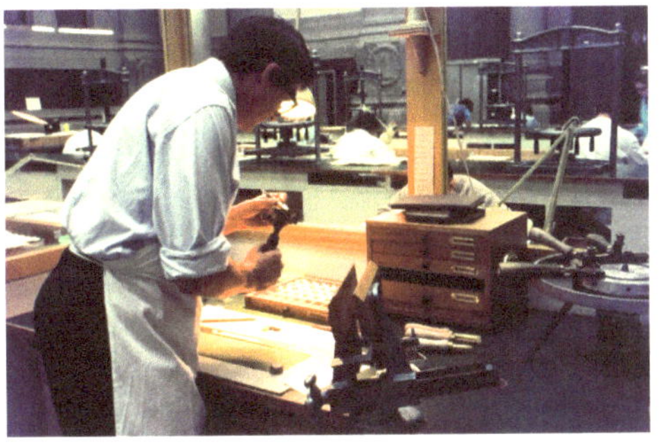

Figures 97, 98, 99 and 100 Covering and finally lettering in gold foil by Roberto Bartolini.

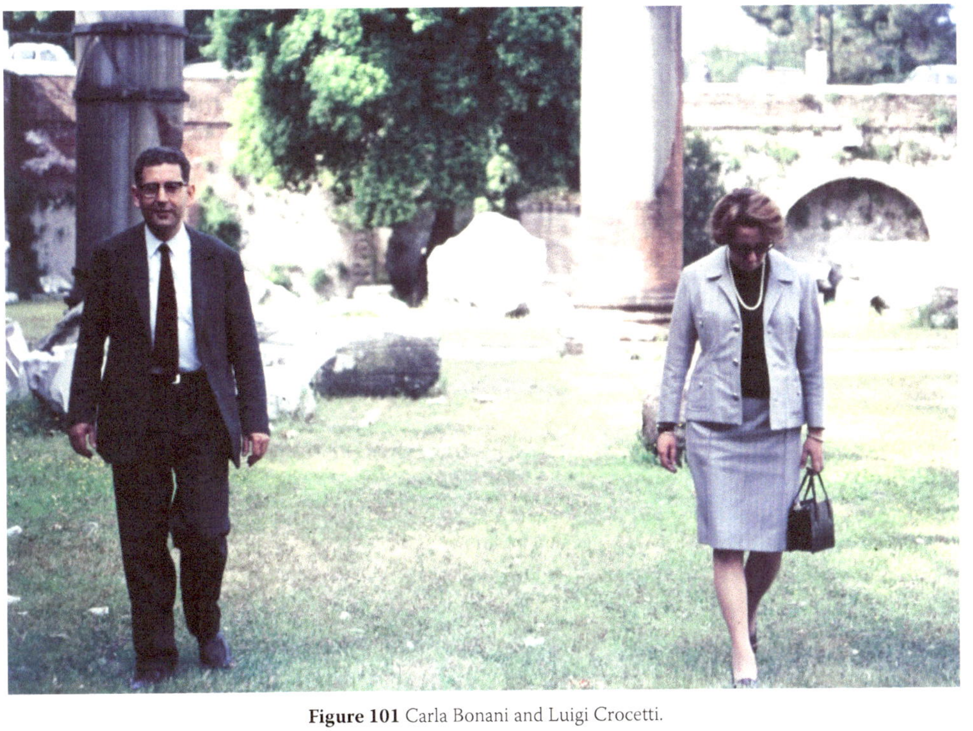

Figure 101 Carla Bonani and Luigi Crocetti.

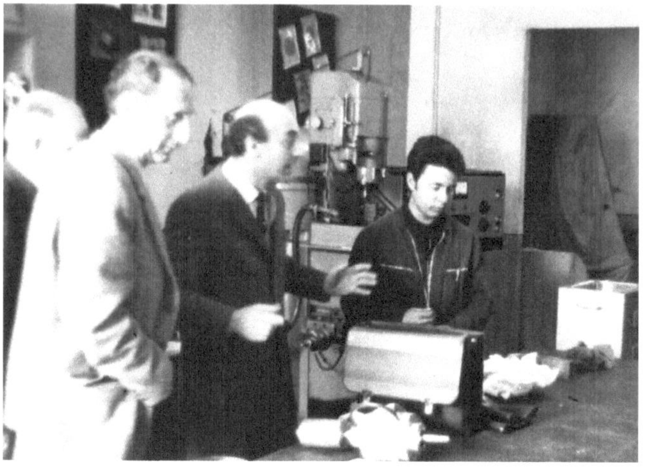

Figure 102 Professor Casamassima in Milan (on the far left).

Bibliography

Anon. (1970) *The Restoration System of the Biblioteca Nazionale Centrale di Firenze*, 2nd edn, revised October 1970 [actually prepared by Dr Crocetti, Joe Nkrumah and Anthony Cains].

Cains, A. (1984) 'Specifications and observations', *New Bookbinder* 4: 61–81.

Cains, A. (1985) 'Specifications and observations', *New Bookbinder* 5: 27–55.

Cains, A. (1992) 'A facing method for leather, paper and membrane', Conference Papers Manchester, Institute of Paper Conservation, 153–7.

Epoca (17 January 1971) 'Rinascono I libri dagli impasti di carta e fange' [see pp. 46–7 for pictures of Joe Nkrumah, Franca Manera and Dr Crocetti to the left and Anne Pelikan and the restored *Teuerdank* to the right].

Maximilian I and Melchior Pfinzing (1517) *Teuerdank*, Hans Schonsberger, Nuremberg in *The Typographic Book 1450–1935*, S. Morison and K. Day. London: Ernest Benn (1963), pl. 49.

Reed, R. (1972) *Ancient Skins, Parchments and Leathers*. London: Seminar Press [In this book Dr Reed misunderstood the term 'limp vellum' in the belief that it referred to the character of the skin rather than the method of bookbinding.]

Training in book conservation after the flood

Christopher Clarkson

As a young man, two years out of the Royal College of Art – taught by Peter Waters, influenced by Roger Powell – I was advised to work for Sandy Cockerell because a bright and skilled young bookbinder, Anthony Cains, was there. Cockerell's father, Douglas Cockerell, was the binder to William Morris. Thus survives a direct line back to the Arts and Crafts movement of the 1900s, with its strong belief in quality craftsmanship and functionality.

Around 19 November 1966 I was contacted and asked to join the British Museum team in Florence. After an intense few days collecting various tools and materials required by the team, I left on 24 November, landing in Milan and taking the train south. It was early evening when I arrived in the main Florence station; due to power cuts, it was pitch black. As the passengers around me dispersed, I remember seeing through the gloom a large grin of white teeth and sparkling eyes. It was my first meeting with Joe Nkrumah.

Some 20 days after the flood, the English were now lodging appropriately in the Pensione Bretagna on the Lungarno Corsini. The next day I was put in charge of volunteers washing and drying books at the railway power station. Professor Emanuele Casamassima, the director of the Biblioteca Nazionale Centrale di Firenze (BNCF), had negotiated its use, as well as railway carriage accommodation for the volunteers and food at the railwaymen's *mensa*. The power station had been cleared of mud and oil, basins and taps had been plumbed-in down the main hall and drying lines set up around the boilers. Generating power was vital and so we had to work harmoniously with the power workers; in those days a dirty, oily business. Our group of lady volunteers so often smoothed tense moments with a well-directed smile.

Some important decisions had already been taken before I arrived:

- The great fear was mould. With the overwhelming number of books, large-scale refrigeration was not possible – in any case there was no electricity. So within the first week, the director decided to force-dry the soaked material. I was told later that over 400 truck loads, almost 3000 tons, were shifted out of Florence to industrial ovens and kilns within the first month (Figs 1 and 2). Another problem was the transfer of dyes (Figs 3–5).

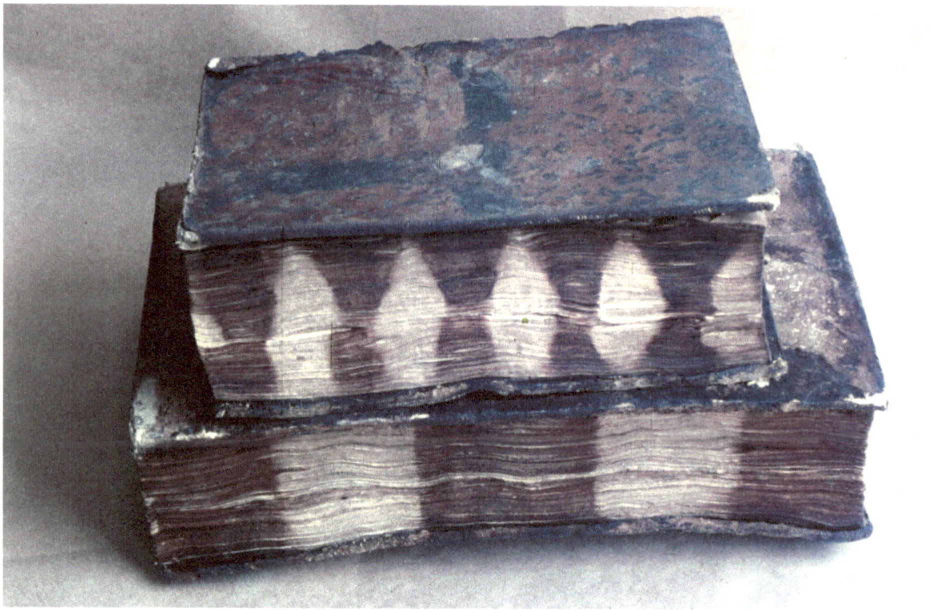

Figure 1 Characteristic striped marks on the edges of books returned from kiln drying. The pattern appears to be a result of the way books were stacked in the kiln: the hot air drew gelatine size to edges causing dark areas, the light areas being protected by racking or by books underneath.

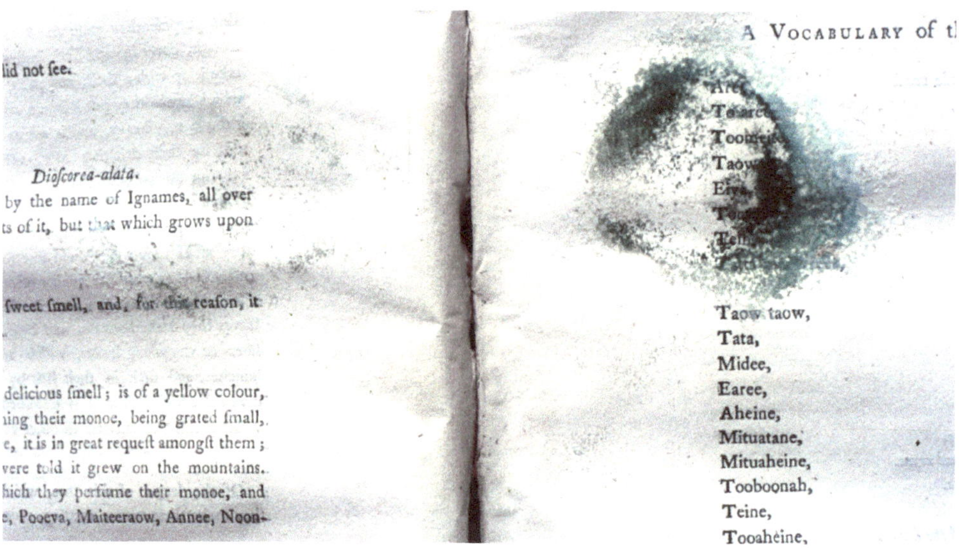

Figure 2 Mould was an early problem, not only for the flooded books, but also for books above the floodwaters due to the effects of high humidity.

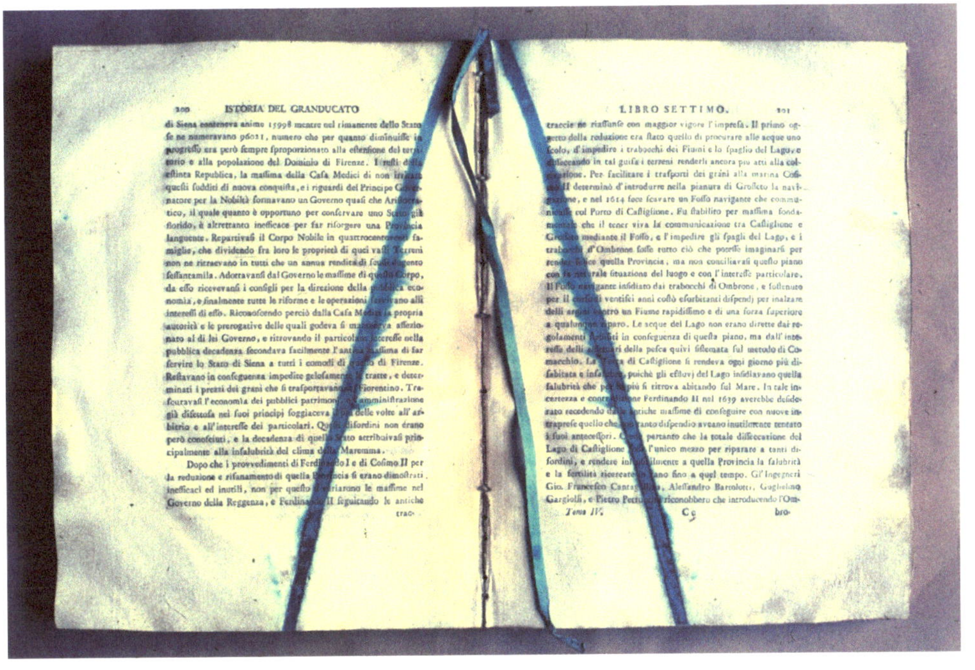

Figure 3 Blue dye from a ribbon marker.

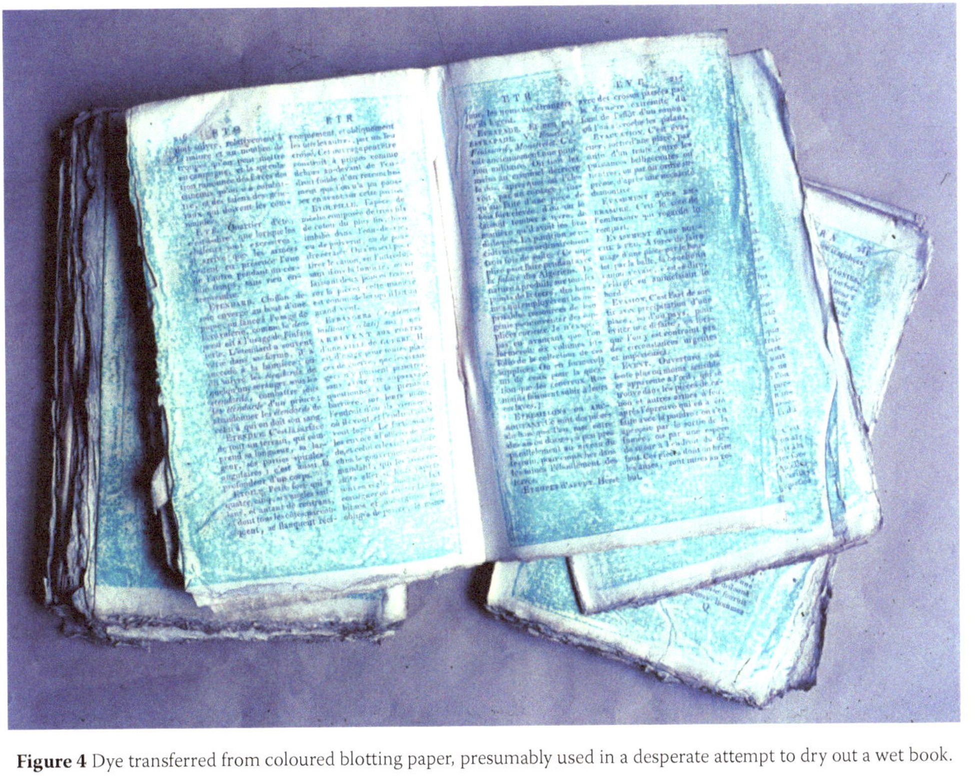

Figure 4 Dye transferred from coloured blotting paper, presumably used in a desperate attempt to dry out a wet book.

- Although a large number of libraries and archives had been overwhelmed by the floodwaters, it had been decided to concentrate British resources on the holdings of the BNCF (Figs 6 and 7). In the first two weeks too much damage had been done by washing, interleaving with inappropriate blotting paper or dismantling books while wet (Fig. 8).
- It was obvious to the English team that conventional techniques could not be applied in the face of such an avalanche of damaged material. It was evident that new approaches had to be devised urgently and so the returning lorries, with their dry but mud-and-oil-caked bundles, were directed to the Forte di Belvedere.

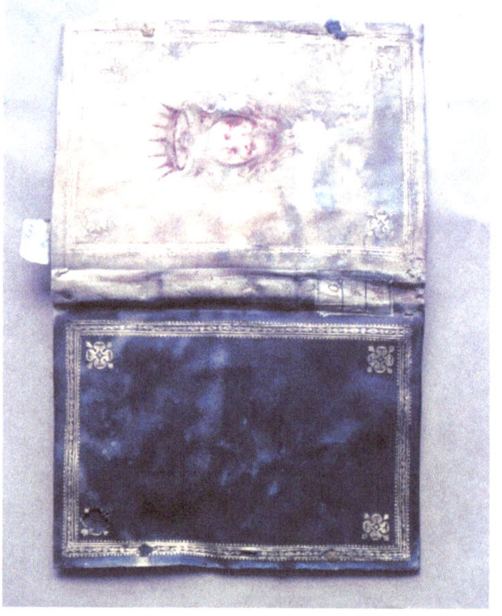

Figure 5 Limp vellum binding with the Medici arms: dye has migrated onto its lower cover from the neighbouring tanned leather.

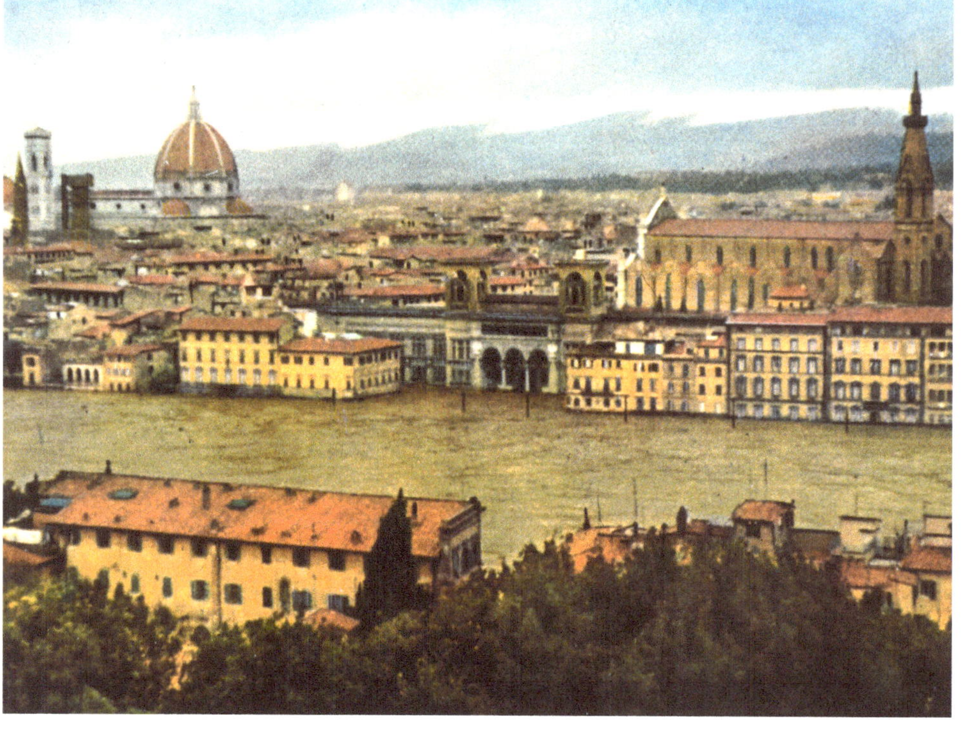

Figure 6 Seen here the morning after the flood, the BNCF in the centre, facing the Arno.

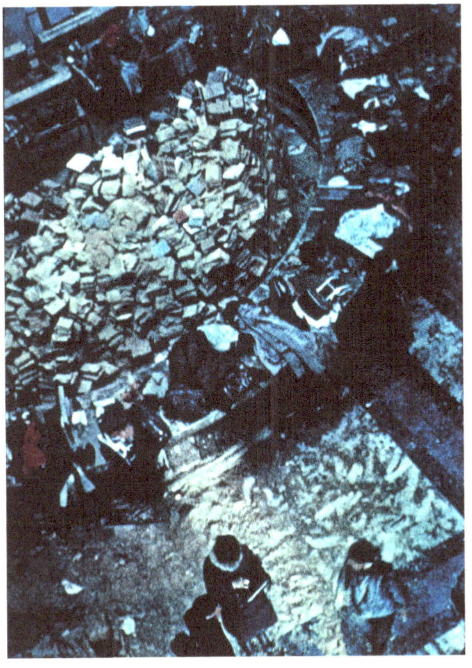

Figure 7 Main entrance to the BNCF. The devastated main catalogue after the waters had subsided.

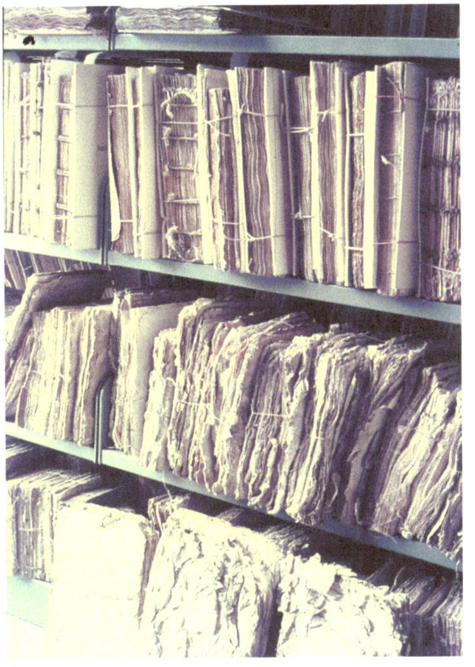

Figure 8 The conventional method of 'pulling' a text block wet did a great deal of damage. Compare the lower shelf of text blocks washed in the less controlled first few weeks with the upper shelf of similarly flood-damaged text blocks that were washed on wooden float boards with the dunking method. Many were washed with their sewing system partly or wholly intact, a change that saved much historical integrity as well as repair costs.

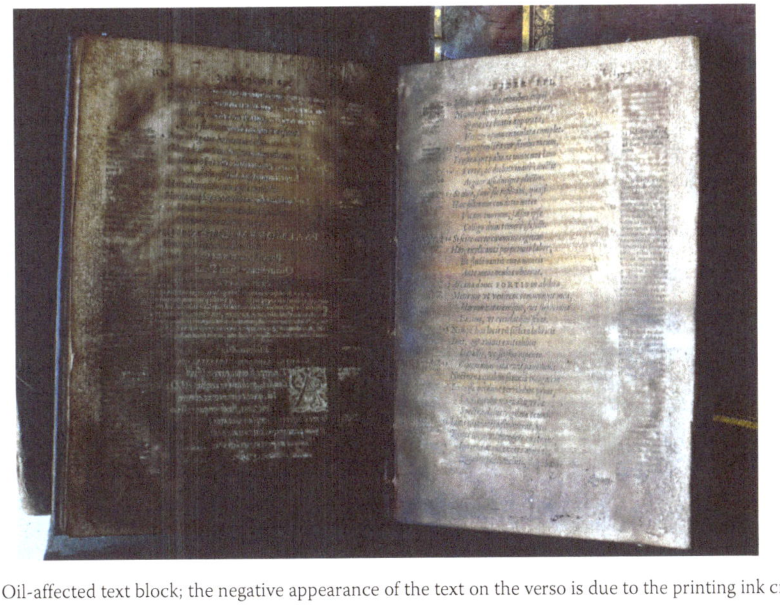

Figure 9 Oil-affected text block; the negative appearance of the text on the verso is due to the printing ink creating a

This *fortezza* had been turned into a central clearing-house where the bundles were sorted and marked into categories of type and extent of damage. Any distorted covers were removed from the more straightforward books and, by careful dry scraping of spines and cutting of sewing systems, signatures were separated from text blocks. Each book was then wrapped and marked with descriptive symbols and, if possible, its shelf-mark. The Fortezza did not have adequate supplies of running water, hence the arrangements with the power station. Books damaged by oil were kept back and experiments carried out with solvents, Fuller's earth etc (Fig. 9).

At the power station I learnt the particular abilities of each volunteer, so that by noting the symbols and checking the contents of each bundle I had a reasonable idea of how to distribute the work. The text blocks were washed, first in water and then in a solution containing a fungicide. The washers' basic technique was to lay a signature on a wooden float-board while holding its spine-fold to get the water to flow into the fore-edge by controlled 'dunking'. This parted the leaves, allowing each folio to be gently washed with a 'soft-hair' brush. The signatures were then gently pressed to squeeze out surplus liquid. Those text blocks still with plate or print impression, or those of poor quality nineteenth/twentieth-century paper, had to be semi-dried with white blotting paper (Fig. 10).

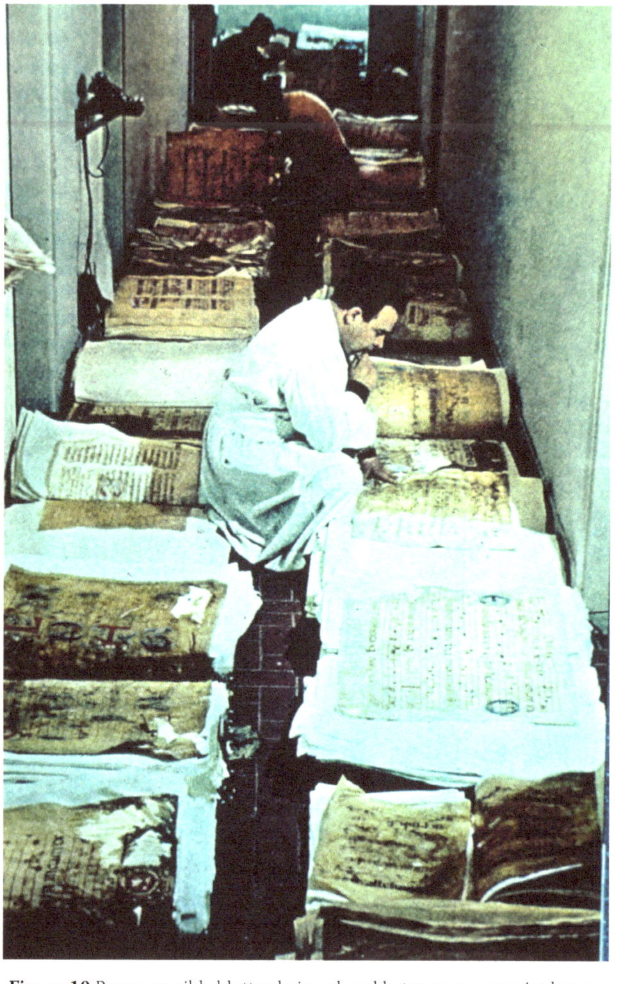

Figure 10 Paper, possibly blotter, being placed between manuscript leaves.

The text blocks were taken to the drying lines set up around the boilers, where each damp bifolia was carefully parted and hung over the lines, order being retained by making sure that the bifolia in each signature slightly overlapped its neighbour. Thus they were dried in the cur-

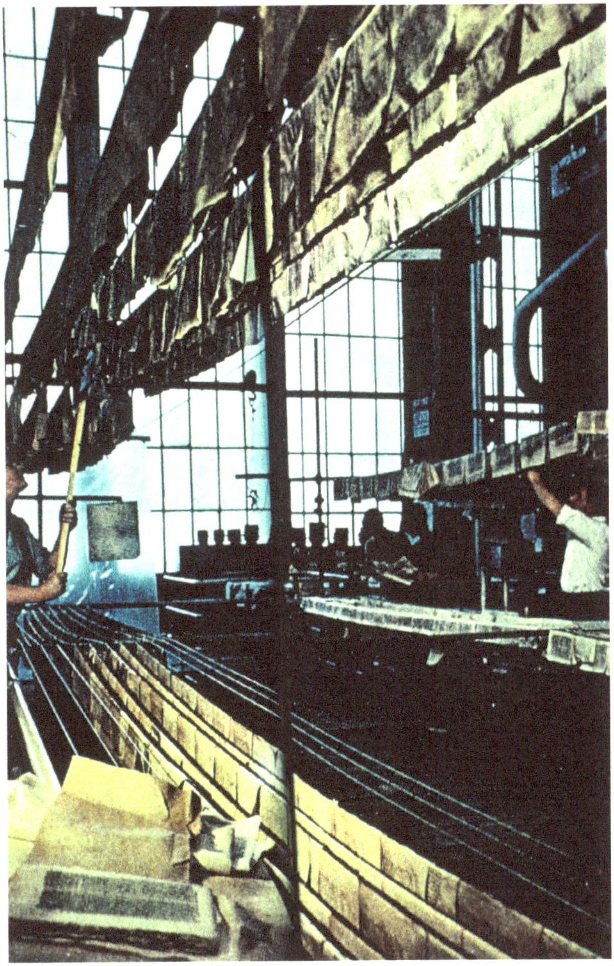

Figure 11 Volunteers in the railway power station, November 1966. A peel is used to lift washed bifolia onto lines.

rents of warm air (Fig. 11). When dry, each signature was gathered and its text block re-formed; it was then taken to a table for checking and collating by a member of the library staff. The text blocks were then boarded and wrapped in a fungicidal paper, inserted into plastic bags, labelled and stored to await the final stage – rebinding.

Of course nothing went so smoothly: I remember a dreadful moment when one of the power station workmen opened a boiler door and the burst of hot air lifted all the bifolia off the lines and they drifted up and up into the high, cathedral-like roof, then glided around on the warm air currents like seagulls, until they gradually descended. We were all running around trying to catch them before they hit the oily machinery, dirty boilers or wet floor. Another incident concerned a television crew who came in and re-spread mud and oil all over the walls, from buckets we had left outside, to make their evening 'news programme more realistic'.

Howard Nixon's (British Museum) official report on his visit (4– 6 December) records that 'The very last books to be removed from the mud were coming up from the basement by human chain to lorries at the front door of the library (BNCF) when I arrived.' He also reported that the rooms above were being cleaned of their residue of slime and being disinfected while new temporary shelving was being installed. He finishes by saying 'One of the great problems is going to be space, as the director has decided not to ever use the basement rooms for book-stacks again.'

The work was piling up at the power station so I continued to work with many of the volunteers through that first Christmas. By the spring of 1967, the British team's whole operation had moved into the BNCF.

An early plan was to send the text blocks out to volunteer workshops – indeed many offers came from governments and text blocks were selected and sent. But due to work quality inconsistencies and worries concerning the more seriously damaged items, it soon became obvious that much of the rebinding work would have to be done at the BNCF. It was decided to utilise the entrance floor and some balcony rooms of the library. By Easter, construction of the workshops and a central materials and equipment store was well on the way (Fig. 12).

Vast numbers of rare and important early books from the sixteenth to the eighteenth century were damaged, comprising the greater part of the Magliabechiana collection and the larger format books in the Palatine library (Fig. 13). Other papers have discussed the design of the workshops, the work of the librarians in attempting to identify catalogue information, the documentation card etc. The language problem for the photographic documentation was overcome by my idea of drawing up a card with all possible views of a book. By circling those required for a particular item, the photographer acquired his instructions. Each book with its specification card was then passed through to a room where, with Tony Cains, we attempted to specify the type of repair and rebinding necessary. I photographed a typical trolley load of damaged books in front of me on Easter morning (Fig. 14).

As the days went by we began to note diversities of structure and their differences along with the range of qualities in materials used, all emphasised by the same cat-

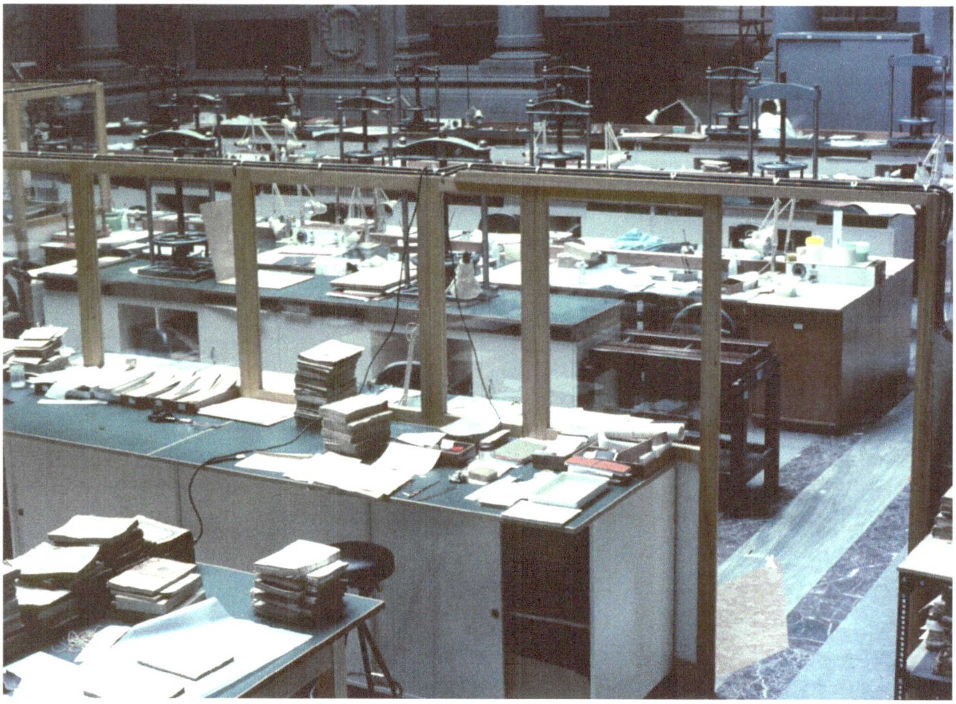

Figure 12 The new workshop utilising the main reading room of the BNCF.

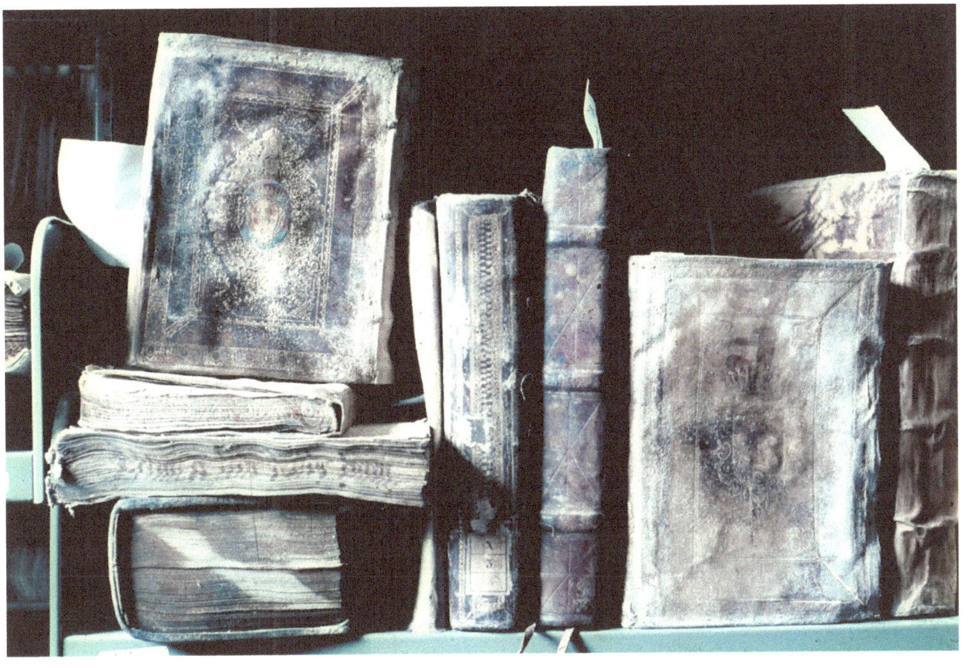

Figure 13 The smaller format books of the Palatine library: note the arms of the Medici. Although above the floodwaters, they were affected by the high humidity. Tannins have migrated from leather coverings to parchment ones.

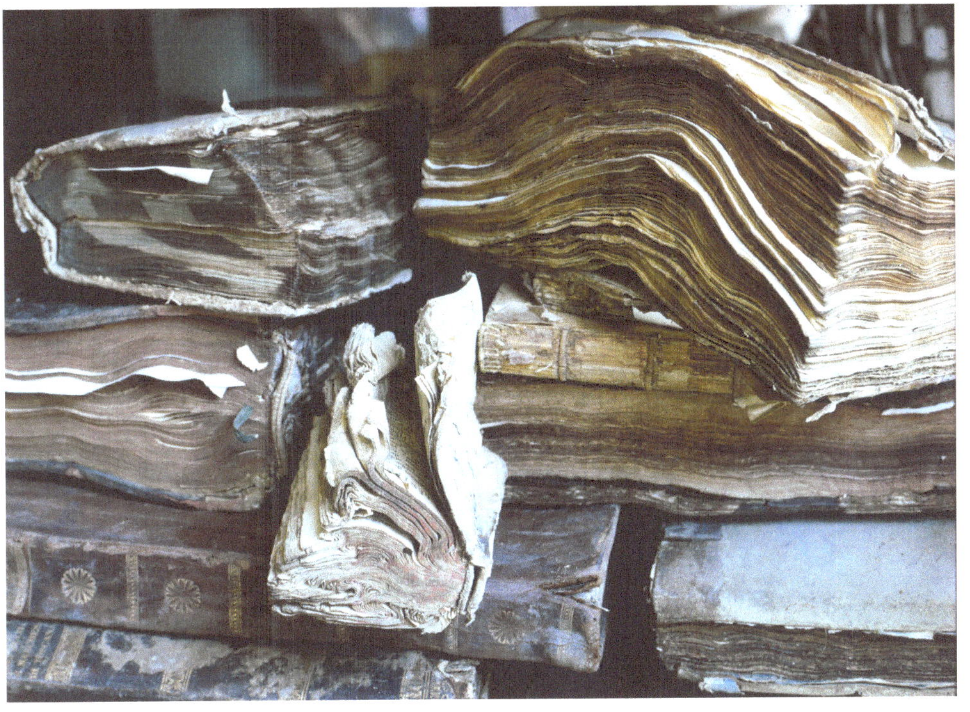

Figure 14 A trolley load of damaged books.

astrophic event. Almost unconsciously, our knowledge expanded and an important study collection built up around us consisting of a wide range of period bindings, in all states of condition (Fig. 15). From all this information, one became aware that what needed to be preserved in a period binding was the binder's underlying intention, the quality of materials and diversity of techniques used.

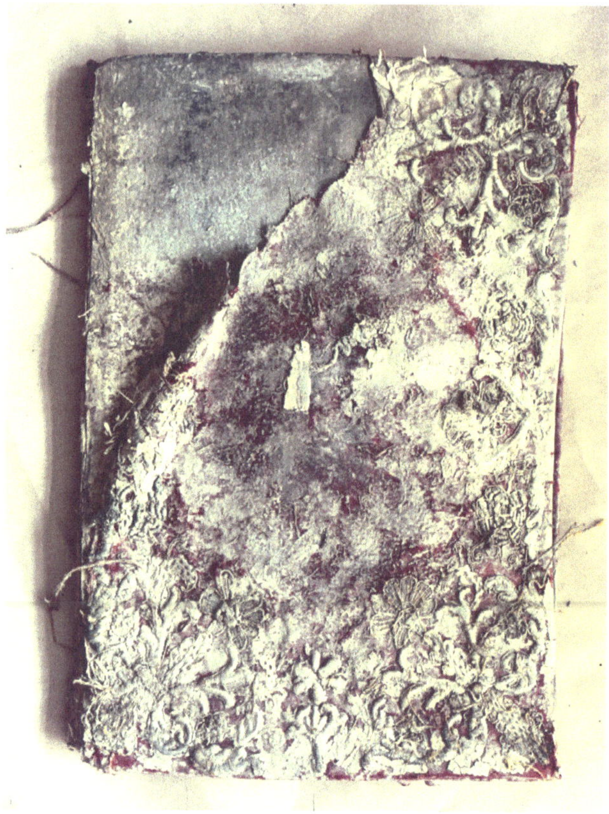

Figure 15 Decorated velvet binding from the Palatine collection.

Personally I had started the year 1967 with the idea that the English craft tradition was a sound base for any training programme and finished the year sensing that it could not cope with many of the problems and the observations that were being made.

As work and my studies in Florence and other libraries progressed, I began to realise that the fifteenth-century form of book evolved from 1300 years of constructional experiment with the codex form of book. The process, which began with the birth of the codex in the first century AD, developed diverse constructional features from the accumulation of different cultural, religious and material influences. By the fourteenth century, there was a wide structural repertoire which, in Europe, only began to codify into a few particular types during the fifteenth and early sixteenth century. By the late nineteenth century, bookbinding was a skilled craft with little soul.

In teaching we started to deliberately apply the word 'conservation' to our activities in summer 1967. Indeed, this was an intentional attempt to show a distinct philosophical break with hand-binding and tradecraft conventions. At that time the word 'conservation' had a certain 'shock' value by association – a 'book as a museum object'? Of course, now the word has been devalued by commercial interests and by institutions that have simply renamed their in-house binderies as 'conservation binderies', with no attempt to rethink, retrain or re-staff. To my mind, the word 'conservation' is also being misapplied when describing certain processes such as deacidification or lamination when, as mass treatments without prior careful selection of items, such processes can be extremely damaging.

'Conservation' has three main strands: the 'craft', the 'chemistry' and, for want of a better phrase, 'historical awareness'. Many young people become fascinated with the craft or the chemistry involved, but developing in them a true sympathy and response to the object as an historical artefact, with all the indefinable qualities that earlier cultures and ageing processes bestow is extremely difficult, but it is an essential ingredient which, by definition, a conservator must have. By this, I do not mean superficial knowledge that turns out those facile facsimiles so often mistaken by the layman as skilled representations of restoration, but to develop a far more sensitive and perceptive person who can, for example, recognise the difference between 'evidential dirt/deposits' and 'institutional dirt/deposits', and understand the reason for saving the former and not necessarily the latter. The 'study collection' we began to build was an aid in trying to achieve this goal and became an essential teaching tool. I later carried this idea to the Library of Congress and later still to the Bodleian Library, Oxford; but as working libraries, such 'study collections' became in reality a list of shelf-marks in 'subject order' that were drawn upon for training purposes.

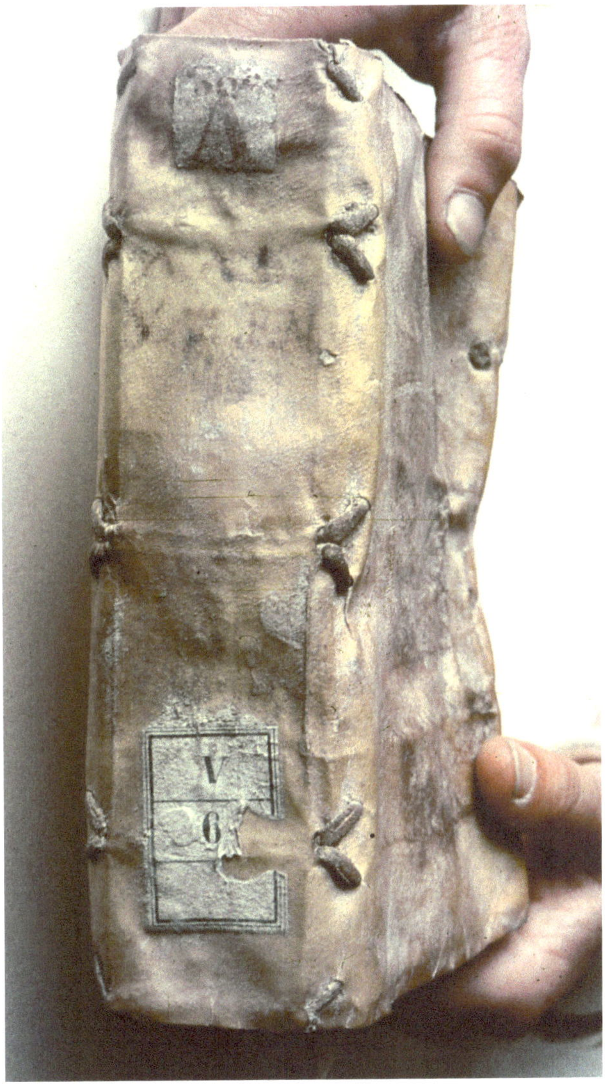

Figure 16 One of the superb early sixteenth-century limp vellum bindings that survived the floodwaters and forced drying to become an exemplar when establishing limp vellum rebinding methods and standards.

It was during 1967 that I became fascinated in the limp and semi-limp paper and vellum bindings typical of a scholar's library, such as Antonio Magliabechi's (Fig. 16). Many of the books had survived the flood and emergency heat drying surprisingly well. In analysing such bindings, I realised that they displayed many conservation principles as well as being good binding structures for training due in part to the

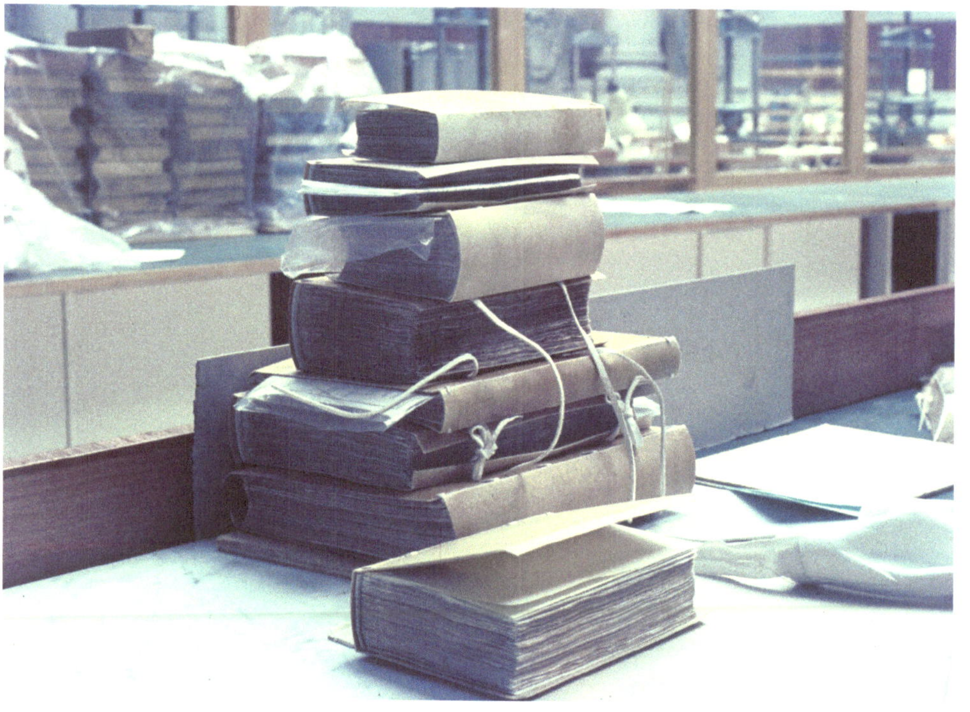

Figure 17 The limp- and semi-limp binding work area: note the specification card for each book.

accuracy required and the lack of the inbuilt tolerances of much stiff board binding, and in part to the necessary harmonies required between the structural detail and the particular quality of the materials used, each being dependent one on the other. Thus they are seemingly simple structures, but they tread a very narrow line between success and failure (Fig. 17).

Searching for quality materials highlighted an important truth that when a particular material, processed by man, is necessary to a society or is 'in common use', then it will gradually be produced with growing sophistication and diversity of qualities all appreciated by and used in that society, but which later generations cannot possibly guess at. In those early days obtaining suitable quality in cover papers and leathers was difficult, but parchment and alum-tawed skin were still being produced with simpler processing, and thus it proved a way to get started.

I contacted Dr Ronald Reed, head of the Procter Department of Food Science, Leeds University, who later came to Florence. Although a modern leather chemist, he was enough of a humanist to appreciate our needs. He was able to supply and discuss possible recipes and medieval techniques which would be useful and why a parchment maker rather than a tanner should be encouraged to produce tawed skins.

It was important for our trainees to realise that it is never possible to accurately record indigenous craft knowledge and empirical skills, so that a scholar attempting to describe how a craftsman achieves a certain effect usually resorts to a hopelessly generalised description. Skin is unique because, depending on the processing, it can supply abrasive resistance and suppleness, incredible strength with softness, directional rigidity with flexibility. It can be elastic, yet tough; tenacious, yet not harmful to a text block. Most important of all, it is capable of being extremely durable.

For limp and semi-limp vellum bindings, two different qualities of parchment were required (for covering and for board-sheets) and three different qualities of alum-tawed skin (for bands, spine patches and fore-edge ties). The sample pieces of vellum shown in Figure 18 are all made of goatskin from the same region in Abyssinia. The difference in skin quality that these two variations alone have produced is quite dramatic, ranging from supple, opaque and extremely strong to, at the other extreme, skins that are hard, transparent, whiten when creased, and even with their outer layer tending to split on folding. Some of the qualities they display I have only handled in vellum skins from certain areas of the medieval world.

In Italy, quality limp and semi-limp vellum binding lasted through until the 1520s, after which generally a decline is observed in the quality of materials and techniques used; by mid-century, it has essentially declined into a 'casing' structure. At its best, it was a non-adhesive structure where joins were flexible, being stitched, tacketted, laced etc., and thus less likely to distort and strain in fluctuating environments.

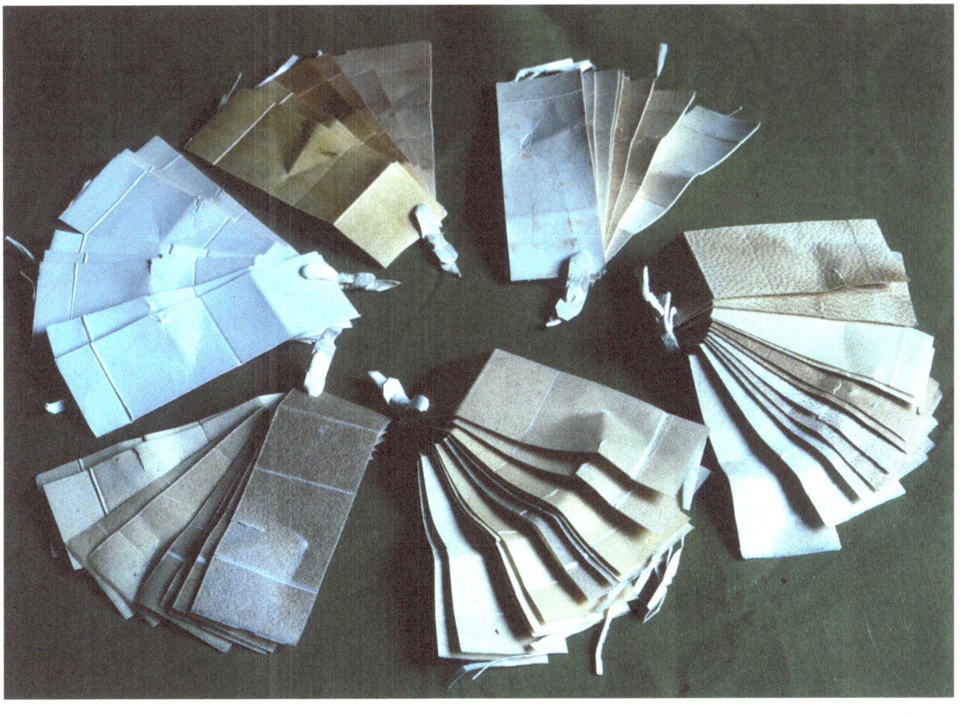

Figure 18 Covering vellum made under the guidance of Dr Ronald Reed, head of the Procter Department, Leeds University. These sample pieces of vellum are all made from goatskin of similar age and from the same region. All the skins were processed using the same recipe, but with varying time in the lime poke and drying times. Their different qualities have all been assessed by my bench test.

Nowadays part of my work involves consulting with library, archive and museum management, which often makes me fear for the future of historic objects. Gone from management are those scholars of the Mechanical Age, such as Professor Casamassima, who understood and respected practical or historical values and the knowledge and skills that made the objects. Is it because the Electronic Age has come about so swiftly that now so much library and curatorial training is theoretical and does not have the large 'workshop' segments it once had? How can one discuss conservation problems with influential people who, in their training, never felt the magic of forming a sheet of paper, printing from lead type or making an etching, let alone a sugar aquatint?

What mass production and theoretical teaching cannot do is to copy or understand the diversity of historic handcrafts. The important idea here is that 'free workmanship' exploits the many different qualities of a natural material. Technique is essentially about rhythm and spontaneity, which in turn is much to do with the imagination and past experiences and skill of a craftsman, qualities in a period object that are of great value to future generations and worth preserving.

So very much was lost in the Florence flood and its aftermath, but I fear much more of historic value will be lost in future disasters. Recently we experienced major power cuts throughout Europe – what will such management do when power cuts are the norm?

Where is the progress?

Joe Nkrumah

In many ways, coming back to Florence has been rather disappointing as I feel that all the efforts we put in, together with the ***angeli del fango*** (mud angels) were in some measure in vain – the job we started has never progressed and is far from its conclusion. Even the idea, born from the efforts and goodwill in Florence, to create an international institution to finish the job in hand was not achieved (see below under heading 'Proposed international centre for the preservation of books and manuscripts').

Following the flood, I stayed in Florence for seven years. I was so enamoured with working with all those rare materials I could not tear myself away. So when I finally broke from Florence and worked in other parts of Europe and then the rest of the world, I always came back to Florence as a kind of coming to Mecca, a pilgrimage. But each time I felt a little bit of me dying because the work was not progressing – more than half of the items 'rescued' from the flood are untouched and still in store. While humanity can find plenty of money and resources to fight wars, little can be found to preserve our art. Culture is regarded as the caboose van – the last one to hitch on and the first one to hitch off.

Coming from Africa, in a way I represent a whole continent – those who were in Florence four decades ago as well as those affected positively by the aftermath of the 'Florence 66' paradigm in the conservation and heritage management world. The excuse for not finishing the job cannot be attributed to lack of means as many ways have been elaborated in the process.

If art is important, if culture is our soul, our identity, our heritage, why do we treat it as if it is ephemeral? We need to teach students methods of preserving collections – not just how to restore and conserve individual objects while the rest languish and decay in the dungeons.

This symposium celebrates 40 years since the flood. Let us now look forward 40 years, with the aim that every shred of paper, every piece of our own heritage that has been put aside in Florence could once again be put on the shelves of museums, libraries and/or in viable storage. Then future generations will know that we were serious and that we have come out of the woods.

The following text is taken from a pamphlet prepared in 1969 by James Lewis, Peter Waters and Professor Casamassima:

Proposed international centre for the preservation of books and manuscripts

> The preservation of library material has become a serious and urgent concern. Although the actual problems vary considerably between countries and individual libraries, they are felt, nevertheless, in all parts of the world and the aid of advanced technology will be needed if the dangers of progressive deterioration are to be avoided in the future.
>
> It is widely recognized that degeneration is due chiefly to air pollution, unsuitable storage conditions, the inherent liability of book materials to deterioration, natural hazards such as fire and water, lack of care in handling and poor restoration. Other less obvious factors, such as unwise preservation policies and practices, which may be irreversible, also play an important role.
>
> It is difficult, moreover, for librarians and archivists burdened as they are, to acquire the knowledge necessary to enable them to make wise decisions about how the material in their care is best to be preserved, to keep themselves informed about practical developments in the field, and to evaluate the qualities of the work of conservators, restorers and binders. If preservation programmes are to be efficiently carried out in the future, conservators with special qualification will be needed.
>
> With notable exceptions, librarians, archivists, restorers and book binders are insufficiently trained in the principles of conservation to cope with the present situation, and training facilities are almost totally inadequate.
>
> It seems essential that book and document restoration should have a status comparable to that of painting and sculpture restoration, and, if scholarly and scientific standards are to be achieved, should eventually attract practitioners of similar quality, drawn largely from post-graduate students.
>
> Within the narrow limits of library and archive budgets, it is difficult to provide money for the financing of training programmes in book conservation. Existing conservation departments in libraries and archives are fully occupied in repairing and conserving their own material. They rarely have the time or the money for research or for training staff from elsewhere.
>
> There appears to be a real need to establish a centre based upon existing departments of conservation and able to profit from its material and experience but devoted to research and training. It is the purpose of this pamphlet to set before you a proposal to establish an international training centre for book and manuscript conservation in an attempt to meet this need.
>
> In the flood of 1966 in Florence, the damage suffered by major collections of manuscripts and printed books from the earliest days to the 20th century was on a far greater scale than that suffered by the art treasures. Generous donations of money

and expertise flowed into Florence from many countries and restoration laboratories were established in the Biblioteca Nazionale Centrale with the help of this foreign aid and with the support of the Italian government.

Because of the unique circumstances, the restoration laboratories had to face each new problem as it arose. New methods had to be devised to cope with conservation and repair on a large scale but related in treatment according to the category of material. As a result, the traditional concepts and practices of book preservation have been subject to reassessment and new criteria have been set up. In effect, a revolution is in progress.

The BNCF operation has brought together a large body of international experts and their experience has been shared. Many restorers sent to help train unskilled personnel have left behind them knowledge, which has contributed to the development of restoration methods, yet have returned home with added awareness and new attitudes of mind. The BNCF is unique in being the only place where a cross-section of almost all the possible problems of book and print restoration are being dealt with, under one roof, using a wide variety of techniques. No solution to any conservation problem is accepted solely on grounds of tradition but everything is subjected to scientific appraisal.

It would now seem natural and desirable to take advantage of this situation and establish a centre for the benefit of libraries throughout the world and also to develop further the training of the permanent Italian staff.

It is, therefore, proposed to set up an independent international organization with an office in Florence, administered by an organizing committee affiliated to the National Centre for the Study of the Preservation and Restoration of Cultural Property in Rome, and in collaboration with the Istituto di Patologia del Libro and the University of Florence. The organization would be established, following a successful appeal for foundation funds.

The Committee would be responsible for the arrangement of courses, the appointment of permanent tutorial staff and visiting lecturers and the selection of applicants for training. The facilities for training will be provided by the Biblioteca Nazionale Centrale.

It is hoped that applications would be forthcoming, not only from students, but also from practising restorers and binders, librarians, archivists and others concerned with the care of books and documents, and comprehensive coursers of varying length would also be arranged to suit their needs and abilities.

The proposed subjects for study, in broad outline, would be as follows:

1. The historical development of book technology, comparative studies and evaluation of book structure. The experimental construction of non-destructive type binding. Composition and testing of book materials and an examination of the testing apparatus. Practical studies of reversible restoration of manuscripts, books, prints and archives. Assessment of mass deacidification projects.

2. The principal causes of deterioration of library and archive materials. Day to day care. Techniques in surveying of collections to establish the priorities of conservation.
3. Emergency measures and after treatment and storage of fire-and-water-damaged materials.
4. The study of the book in relation to conservation practice in so far as it affects the scholar, scientist and restorer.
5. The administration of preservation programmes in libraries.

Questionnaire

We should be grateful if you would reply to the questionnaire below, adding any other comments you feel appropriate:

a. Do you approve and welcome these proposals outlined above?
b. Could you make use of the training offered?
c. What would be the qualification and experience of your prospective candidates?
    - Librarian
    - Archivist
    - Conservator
    - Book restorer
    - Document repairer
    - Binder
    - Post-graduate student
d. If you already have restoration facilities would you be prepared to give some training in your speciality to one or more BNCF personnel on an exchange basis paid for by an international fund?

The Directors: James Lewis and Peter Waters, Conservation of Library Materials Research Group, Imperial College of Science and Technology, Exhibition Road, London SW7, England and Professor E. Casamassima, Director, Biblioteca Nazionale Centrale, Firenze, Italy.

Improvements in the treatment of individual books as a result of the flood: a personal review

Dag-Ernst Petersen

Introduction

When considering the history of book conservation within the last 50 years in Germany, it is striking that two disasters proved to be milestones in the trend towards improvements in this field. Following the Second World War, the foundation of methodical book restoration was laid out by the Bavarian State Library in Munich. The collections of this outstanding library suffered disastrous destruction during the war, and after 1945, the library established a well-equipped laboratory for book conservation. All treatment options, both for mass treatment and individual restoration, had to be developed by the staff. Similarly, the Florence flood of 1966 had a far-ranging impact on, at that time, this young profession. The real contribution of the Biblioteca Nazionale Centrale di Firenze (BNCF) was to raise the status of book conservation to an international level. The consequences of another disaster – the fire in Weimar in 2004 – mean that yet again, conservators have to meet new challenges.

In order to gain a better understanding of my comments on the legacies of the flood, some background information about my conservation career up to 1967 (having completed my apprenticeship in bookbinding) might be helpful. Following the war, there was a general demand for restorers to deal with the many different German collections that had suffered damage. Training in book restoration could only be undertaken at the Bavarian State Library, which therefore became the centre for conservation in West Germany for many years.

In 1965, during an internship of six months at the Institute for Book Restoration in Munich, I acquired an extensive knowledge of book and paper restoration. Under the supervision of Butz, Jäckel and Ziegler and their colleagues, I treated medieval books and the corresponding materials, which were either conserved or renewed based on learning by experience. High standards in workmanship were developed and applied in the Munich laboratory; the standards were therefore set and are still universally valid today.

In Austria a similar centre had been set up: the 'Institute for Conservation' at the National Library in Vienna. The head of the laboratory, Otto Wächter, had completed an advanced training at the famous Instituto del Patalogia del Libro Alfonso Gallo in Rome. In Vienna, besides book restoration, emphasis was placed on the treatment of works of art on paper and globes. As I was very curious about these special subjects, I applied and was accepted for an internship.

In 1966, it became clear to me in Vienna that the basic principle of 'reaching one's target by other means' also applies to restoration. In Munich, potassium permanganate for bleaching was favoured, but chlorine bleaching was strictly prohibited; in Vienna the bleaching took place with hydrogen peroxide or chlorine, while potassium permanganate was condemned. In Munich papers were resized with rabbitskin glue solution, while in Vienna, methylcellulose was applied. There seemed to be no communication between these two institutions and no satisfactory reasons for the differences were given.

The 1966 flood in Florence attracted national and international support. The Volkswagen Foundation, on behalf of the German Federal Republic, financed the provision of a number of bookbinders although specific financial details from the Foundation's archive are not available. Accompanied by my wife, I went to Florence in 1967 and 1969 as part of this programme.

Working at the BNCF

Due to my former knowledge and skill in book and paper conservation, a high value object was entrusted to me: the *Teuerdank*, an early printing of 1517 on paper (see Figs 1–5). The originally intensively coloured woodcuts showed characteristic water damage. The colours had bled a great deal and spread over the paper, particularly the green.

Together with the item – already disassembled – I received the standardised documentation card that was used by the librarian and the head of conservation to define the processing steps. There were frequent opportunities to discuss different opinions on operation details with the chemist, Dr Nkrumah, and the conservator, Anthony Cains. During these forthright and frank discussions, I extended my knowledge and was thereby able to contribute further to the restoration process in Florence. This type of dialogue is by far the best way to develop a comprehensible conservation plan and is a central feature of present-day scientific conservation.

In addition, I availed myself of the opportunity to exchange experiences with international experts in conservation, mainly from the UK and the USA, who visited the Florence centre on the occasion of the flood. This atmosphere provided a new impetus and some of the lessons learned can be used as principles for modern conservation.

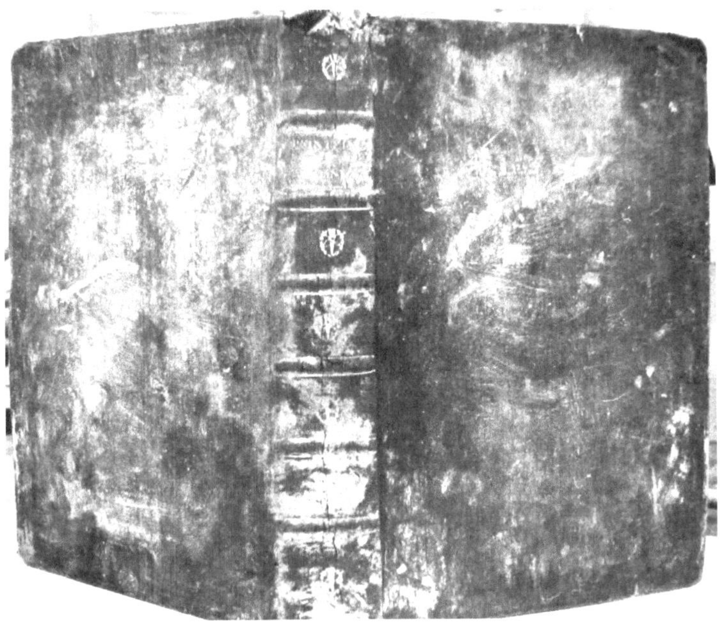

Figure 1 The *Teuerdank*: the old binding after the flood (photo: Anthony Cains).

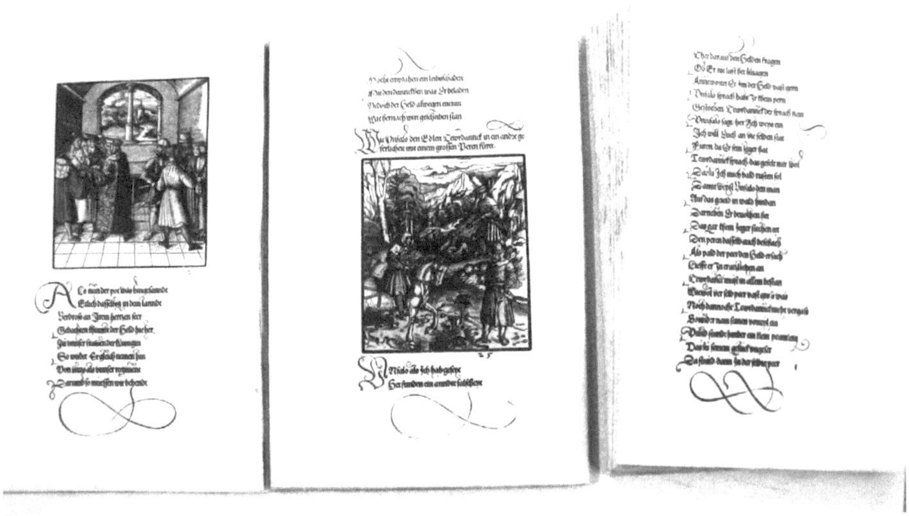

Figure 2 The *Teuerdank*: the pages following restoration (photo: Anthony Cains).

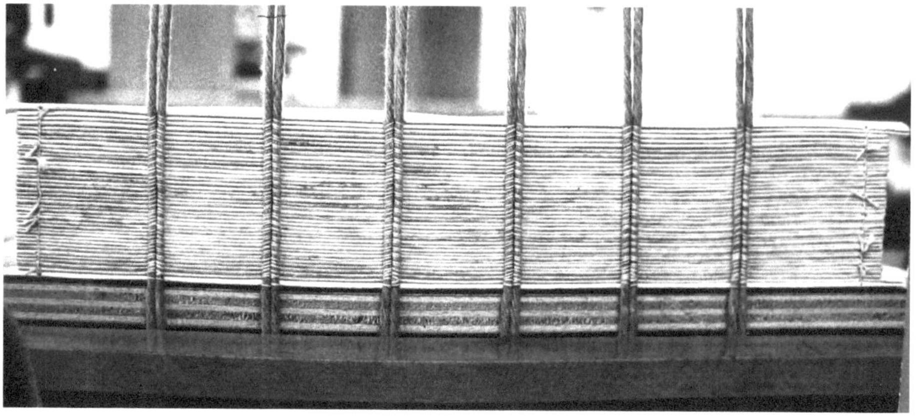

Figure 3 The *Teuerdank*: the book block in the sewing frame (photo: Anthony Cains).

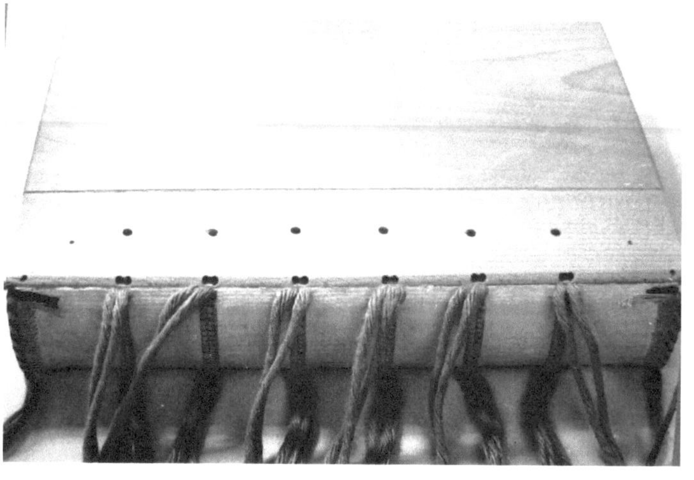

Figure 4 The *Teuerdank*: the wooden boards (photo: Anthony Cains).

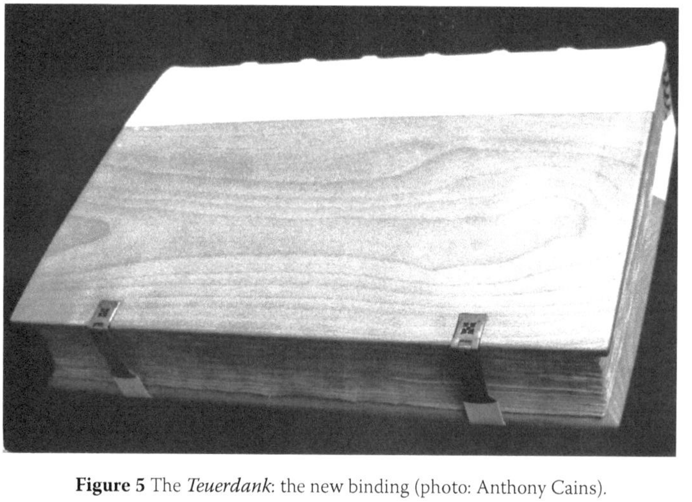

Figure 5 The *Teuerdank*: the new binding (photo: Anthony Cains).

Tools or auxiliary materials

The spread of information about tools or auxiliary materials plays an important role in the refinement of practical restoration – often what may appear to be petty details provide both efficient functioning and increased quality. One example is the tacking iron (a modified soldering iron), commonly used today in paper repair. This implement was employed successfully in Wolfenbüttel, for example on medieval bibles that were not taken apart but required the continuous repair of heavily damaged pages.

Quality standards

Constant consideration of quality standards must, in my opinion, be given to all materials used in book conservation – paper, leather, parchment, textiles, adhesives, plastics, etc. Even though a bookbinder may have completed traditional training several decades ago, he should not be asked to completely avoid the use of materials he became acquainted with during his apprenticeship. During the nineteenth and twentieth centuries, the quality of these materials became considerably poorer, for example endpapers, shirting, batiste and calf leather. The idea of re-restoring a binding in the foreseeable future because a generation of restorers used non-durable material is not an impressive one. There is a tendency for conservators to concentrate too much on the past instead of the future of the objects. In the case of books, they must be able to withstand intensive use in public libraries.

In the Florence conservation laboratory, the qualities of the required materials were defined before they were purchased, mainly in Italy. The same concept was continued successfully, after the flood, by English and Irish conservators who together requested the manufacturer to produce a material according to certain quality standards. In return, they committed to placing a bulk order, which was then shared between them. In Germany, however, it was incredibly difficult to convince one dealer to include materials such as alum-tawed skins, aero linen and linen cord in his range of goods for sale – despite their incredible qualities, he was very anxious as to whether German conservators would be interested in obtaining these products.

Evaluation of chemicals/plastic used in paper conservation

The critical and rigorous evaluation of any chemical or plastic used in paper conservation is paramount. A long list of chemicals recommended to restorers in the 1960s and 1970s can be compiled. Even though the prescriptions were followed carefully, restorers were unable to predict the chemistry of the reactions, the long-term effect on cellulose or the possible damage that might be caused to both their own health and that of the environment.

In Munich and Vienna respectively, I became familiar with chemicals such as potassium permanganate and chlorine for bleaching, Gardinol or FEWA as wetting agents, Nipagin as a mould-inhibiting preservative, and the so-called *Dokumentenlack* (polyacrylate), which was used for repair in Munich. In the Print Restoration Department of the BNCF, technical instructions or 'Procedures to be adopted' were set out (compiled, I believe, by William Bousted from Australia). In these instructions, other materials including mostly English products such as Lissapol, Topane, PVA or Calaton were specified. The procedure worksheet can be summarised as follows:

1. Examination under UV lamp and microscope. Testing with pH meter.
2. Dry cleaning.
3. If oil stained, stains should be removed.
4. Washing in 1% Lissapol solution.
5. Removal of mould, leather and water stains by bleaching with the advice 'always remember that indiscriminate bleaching can cause irreparable damage to paper fibres'.
6. Neutralisation of the chlorine, testing with potassium iodide starch indicator paper.
7. Washing for several hours.
8. Drying by air or heat.
9. Repair.
10. Deacidification with magnesium or calcium bicarbonate by spraying or immersion.
11. Sizing with gelatine or more often with soluble nylon.

Treatment of the *Teuerdank* followed this procedure except that methylcellulose was spread on the pages instead of immersing them in soluble nylon. Current state-of-the art practice in conservation is to use as few chemicals as possible.

Since the flood, there has been an increasing interest in ancient binding structures, particularly with regard to their stability. A survey of the items damaged during the flood showed clearly that certain vellum bindings from the sixteenth and seventeenth centuries survived the water much better than leather bindings of the same period. This experience probably encouraged Chris Clarkson in his further research and experiments with limp-vellum bindings. In the 1970s, there was a lively interest in limp-vellum bindings – especially in the *Koperte* – by English and American conservators, who took the time to visit German and European libraries holding late medieval stocks in order to study the originals.

The general insights into binding structures led to an appreciation of the function of the techniques and their advantages and disadvantages. New understanding was gained of the durability of the materials used on these books. The highlight in this field of research is the publication *The Archaeology of Medieval Bookbinding* by

J.A. Szirmai (Aldershot, Ashgate Publishing, 1999). This research on binding structures in Germany resulted in, for example, the discussion about the so-called 'conservation binding' advocated by Szirmai. However, it is still necessary to integrate the findings of Szirmai, Clarkson and Middleton into conservation practice; if this could be done, an interesting curriculum for book conservation training might develop.

As previously mentioned, most German book conservators received training in contemporary bookbinding. In England this craft had a similar tradition, but in Germany for some time now I have observed a reduction in the quality of the apprenticeships in hand bookbinding, an observation confirmed by the conclusions of ENBOTRAIN 2001 (ENhanced BOokbinding TRAINing in Europe), a European Community project, i.e. that the tradition of craftsmanship might be lost. This will undoubtedly impact negatively on the practice of book restoration.

Conservation implies more than applied chemistry, art history and documentation; boxing of books and documents should not become the only option. Setting up secondary school education systems all over Europe and the USA should be a logical necessity but there seems little possibility of teaching the students 'workmanship'. Preservation today is a task assigned to, among others, conservators, natural scientists, technicians and librarians. Confidence and appreciation of the professional skills involved are the basis of the cooperation.

Conclusion

Conservators are confronted with the all-embracing task of preservation – many of the collections in public libraries are in need of help. The causes of damages are manifold but include natural ageing of the materials, poor storage conditions, acidic papers, intensive use and natural or man-made disasters. With regard to the large-scale needs, it would seem logical to speak of 'mass treatment'. However, in Germany, this term only applies to the treatment of a large quantity of objects having identical damage, for example 'mass deacidification'.

Among conservators there is agreement that a full restoration should remain the ultimate alternative. But if it is decided to restore an object with all the consequences this entails, it is important to adopt the following standards:

- Written documentation complete with photographs to maintain information about the untreated object.
- Exchange of experience and information with colleagues.
- High standards of workmanship and the application of good materials in order to ensure long-term preservation of the restored book.

- A limit on the use of toxic chemicals or chemicals with uncertain/unknown long-term effects in book and paper treatments.
- The application of information obtained by research into the history of binding to benefit hands-on conservation treatments.

In essence, I suggest that some of the fundamentals of present-day conservation are the direct legacies of the Florence flood.

The Florence flood in the popular imagination

Joanna Hines

In November 1966, I had just sat the exams for Oxford and Cambridge. Somerville College, Oxford offered me a place to read History, starting in October 1967, which left me with the luxury of many months to fill. In January, I set off for Italy to help with the rescue effort, one of the hundreds of young people from all over the world who participated in that unique enterprise.

As I was fresh out of boarding school, it was likely that any destination that January would have made an indelible impression, but it was my great good fortune to choose Florence at a unique and fascinating time. I stayed a couple of nights in Rome with the Rome correspondent of *The Times* (my father was foreign editor) and then travelled by train to Florence. My first assignment, however, was not in Florence at all, but with the Contessa Guicciardini at the Castello di Poppiano. She was wonderfully kind and welcoming to the odd assortment of foreigners who had turned up to work on the documents that had been stored in the basement of her *palazzo* beside the Arno, and which had therefore borne the full brunt of the flood. I have a photograph of her sitting beside an impressive-looking radiator, but sadly it was for ornament only. Central heating had been installed in the *castello* some years before at vast expense, worked briefly and then packed up, never to warm the place again. I have never been so cold. It was probably only the liberal amounts of the estate's delicious *vin santo* which was served mid-morning instead of coffee that saved us from pneumonia. The whole place was incredibly beautiful, traces of snow still lying in the hollows – remote and quiet and entire unto itself.

I spent a week there before heading back to Florence and the equally cold and not at all beautiful student hostel in the Via Ghibellina. I was assigned to a team organised by the Victoria and Albert Museum in London under Kenneth Hempel. Over the next two months I was lucky enough to work on sculptures and the occasional gilt frame in the Uffizi and the Bargello. The high point of my career as a conservator came when we were put to work throwing talcum powder at the walls of the Baptistery to draw out the oil and muck that had been absorbed by the stone when the floodwaters poured through. By that time we were being paid by the Uffizi – in theory at least – which probably makes me one of the few professional talcum powder throwers in the world!

It was an extraordinary time. Florence like Florence has never been seen before or since. No cars, no tourists, few shops and restaurants. Mud everywhere. Those inhabitants who could had mostly left, so that in some strange way the city seemed to belong to us. Or so we thought. In March I returned to England – London in the summer of 1967, a good place to be. That brief time in Florence became a kind of island memory, disconnected from anything before or after.

Fast forward more than 30 years. I'm living in Cornwall and have seven novels to my name. I haven't been back to Florence in more than a quarter of a century, but images, memories, themes from that strange time are beginning to accrue, become insistent, not go away – that strange process which is for me the genesis of a novel. The protagonist of this novel is to be a conservator and its working title *Angels of the Flood*.

It is the only book I have ever written that drew directly on my own experience – not the macabre plot, thank heavens, but the description of Florence in early 1967 is how I remember it. However, after 30 years, memories become fossilised as a few vivid snapshots, and I needed some way to free them up. I wanted to recapture the smell and taste and feel of those weeks. My researches threw up some fascinating descriptions on the flood itself.

Newspaper accounts provided the information which I must have read, the images of destruction that lured so many youngsters out there to help. And they provided a context, too. It's strange how much a 'different country' even my own past has now become. In November 1966 the Wilson government had imposed sanctions on Smith's Rhodesia. Dr Verwoerd was assassinated in South Africa. The Red Guard were terrorising Peking (not Beijing). Abortion had only just been legalised in the UK and was still a criminal act in Italy. In the US, Lyndon Johnson was becoming increasingly unpopular.

It's the details that I had forgotten. A fashion writer, travelling on the underground on a hot August day, noted that of 17 women in the carriage, 'only' three were wearing gloves, which until recently had been *de rigeur* for smartly dressed women travelling in public. Fifteen boys who absconded from their public (private) school were found and returned and given 'six of the best' by the headmaster. On the day the flood was first reported, the *Daily Mail* noted that Prince Philip and five companions had bagged 247 pheasants in the first shoot of the season at Sandringham. A vanished world, indeed.

I managed to track down the film Zeffirelli made to publicise the catastrophe and help raise funds. This is a fascinating film, not just for the footage of the flood and its aftermath but also for the light it sheds on the age-old question of what should one rescue from the burning building: the baby or the Vermeer? The Florence flood is that rare beast, an international catastrophe involving very little loss of life. I find it hard to think of another. As a newspaperman's daughter I had been brought up to mock the famously most un-attention-grabbing headline: 'Small earthquake in Ecuador. No one killed.' Only a relatively few were killed in the Florence flood. And yet …

Zeffirelli's documentary is fronted by Richard Burton, who, as well as being Elizabeth Taylor's husband twice over, was also a Welshman. Less than three weeks before the Arno burst its banks disaster had struck a small Welsh village, not all that far from where Burton was brought up. A coal tip slid onto the village school smothering 144 people of whom 116 were children. In a couple of minutes a community lost half its children. And yet, at some stage in the documentary, Burton states quite plainly that in his opinion the destruction of Florence is a greater catastrophe than Aberfan. So he would rescue the Vermeer. Perhaps that's what we would all do – can we remember how many people died in the earthquake in Pakistan in the autumn of 2005? Was it 8000? 20,000? 40 or 80,000?

While hunting down those details that would revive the memories, I contacted a friend from those weeks and asked her what she could remember. She sent a brilliantly evocative description:

> The smells of fuel, sewage, mud, damp plaster.
> Roads dug up – exposed pipes, boards.
> Beach Boys on the juke box.
> Unclear air, dampness, chill. Halo round street lights. Mist in mornings. Haze even in sunlight – contrast to Fiesole.
> Elegantly dressed hookers – French, someone said.
> At night the quiet and the emptiness of narrow streets – echo – loudness of scooters and voices. Mournful. Oily tidemarks.
> People obliged to live on upper floors – tourist areas fixed.
> Chest infections, irritated skin, trial and error cleaning methods, useless polythene gloves – better ones to be shared, then a glut.
> People crossing themselves with holy water scummy white from powder being chucked at Baptistery walls.
> Sound of phlegm hitting the courtyard of the beautiful dosshouse as the old men hawked and spat from upper floors in the morning.
> Skewered uccellini.
>
> *Angeli del fango.*

Hardly surprisingly, that letter brought back vivid memories.

In the Public Record Office I came across a detailed picture of Florence in the hours after the flood, written by the British consul to the ambassador in Rome. In it he described how 'the famous panorama of the city from the Piazzale Michelangelo' had been transformed into a 'heartrending spectacle'. The roads on either side of the river had vanished under the water and a 'strange ripple across the torrent' was the only sign of the Ponte alle Grazie. The river was pouring through the remnants of the shops on the Ponte Vecchio and the 'inland sea around the Biblioteca Nazionale gave some indication of the extent of the disaster that has overwhelmed some of the finest libraries in the world'.

There were, however, as always happens in great tragedies, moments of bizarre humour. The consul described his first contact with a 'survivor', the appropriately named

Reverend Church, who was chaplain in St Mark's English Church (only in real life can you have a chaplain called the Reverend Church – you'd never get away with it in a novel). Clad in fisherman's waders, and aided by his wife, he was digging mud from the vestry and throwing it into the street. Apparently when the waters first poured into his church, the resourceful chaplain had inflated a rubber dinghy and had thus been able to rescue vital records and some of the most valuable ornaments. The consul commented that the chaplain 'nearly suffered personal injury when the reredos became buoyant and joined the floating pulpit and lectern in the swirling waters'. That image of the Reverend Church obliged to take avoiding action to escape a nasty buffeting by the renegade reredos provided a welcome glimpse of farce amidst the horror and destruction.

Research is a wonderful displacement activity when pondering a novel. It fills the time and stops me from launching into the actual writing too quickly. I was extremely lucky in that my closest friend at Somerville, with whom I shared a flat for two years, was Caroline Villers. By the time I was researching *Angels*, she had become head of conservation at the Courtauld Institute in London. She was enormously helpful, naturally. *Angels of the Flood*, which came out in April 2004 was dedicated to her. In September of that year she was diagnosed with a brain tumour and on Christmas Eve she died. All those who knew her know what a loss that has been.

The journey of this book has shown me what a surprising amount novelists and conservators have in common. Those who work in the art world are often selected as characters in fiction, partly because novelists generally prefer their subjects to be involved in creative activity. It is no doubt easier for us to engage with a character who is a painter than say a biochemist or a dental technician. Many writers wish they were artists anyway and vice versa. Even their detectives – another hugely overrepresented profession – often have a penchant for the arts, rather than hobbies such as ten pin bowling or taxidermy.

So it is inevitable that conservators are more common in literature than in life, especially in crime fiction. The craft of the conservator contains obvious elements of detective work – all those X-ray images showing the stages of a picture's evolution. But also, both novelists and conservators operate at the point where reality and illusion intersect. Novelists have a perennial fascination – and not just because they make for useful plot devices – with the whole twilight art world of fakes, forgeries and false attribution. One reason for this may be that a successful novel is itself a sleight of hand: make-believe masquerading as real life. A novelist friend resolutely maintains that she makes her living by telling lies.

In my own novels it is the human fakes and forgeries that take centre stage. While researching *Angels*, I was reminded that the first 'human forgery' I came across was a fellow flood worker. An American called Steve, he showed us photographs of his family in the mid-West – their home, his father's car, even their dog. It was only after a few weeks that we gradually realised he was an Italian. The Steve whose identity he had 'borrowed' in order to escape the draft had been an American friend who had died. Or that's what I believed at the time that I left Florence. Who knows what the truth was? Once the fictions begin, nothing is ever certain.

Reality and illusion. It's the business of a storyteller to devise fictional characters who are entirely credible. Lifelike, in fact. Similarly, I assume, some aspects of conservation are about cheating time, creating the impression for the casual observer – without, of course, any intent to delude – that a damaged or time-worn piece of art is near enough to how it started out. We research, we amass detail, we fret about process and technique, we experiment and try different methods. And then finally, we launch our fictions into the world, and hope the public will be beguiled by our meticulously crafted illusions.

The Florence flood: some personal recollections

Marco Grassi

Over the years, I have particularly treasured two images as precious souvenirs of my early career in conservation. In the first image, I am at the centre, flanked by Thomas Schneider, Anna Brigida and Professor Augusto Vermehren in the old professor's studio at the Accademia in Florence. The year was 1959 and both Thomas and I, as beginner apprentices, divided our time between the main Uffizi Gabinetto, at that time across from the gallery entrance, and Vermehren's much smaller facility at the Accademia. The second photo shows us again, Thomas (standing, right) and me (from the back, left), applying papers to Vasari's gigantic *Last Supper* in the refectory of Santa Croce. The date is, of course, early November 1966, in the aftermath of the catastrophic flood that inundated the city and most of the Arno basin. Both Thomas and I had, in the intervening years, spent time in Rome at the Istituto Centrale del Restauro (ICR), and then, he in Stuttgart and I at the SKI in Zurich. Also, since 1963, we had joined forces in a private conservation practice in Florence. In 1970 we went our separate ways but have remained friends and colleagues ever since.

Learning the profession in Italy in those years – 1958–1966 – proved to be a remarkable experience. Florence was still very much the cradle of a traditional craft that had evolved little since the days of Secco-Suardo and Forni. To be sure, there were highly gifted practitioners such as Leonetto Tintori who were well aware of advances in the areas of materials and diagnostics. A letter dated 1958 from Tintori is evidence of his early interest in newly developed synthetic substances. Even before the Second World War, Vermehren had carried out research in chemical analysis and X-ray technology. But, from a Florentine perspective at least, it seemed as if progress in all these areas was proceeding far more boldly and rapidly in Rome. Indeed, the ICR, under the brilliant and irrepressible Cesare Brandi, had become the forward outpost of experimentation and innovation, an achievement that was due as much to Brandi's remarkable stature as a theorist as to the formidable practical talents of the specialists he was able to attract to the institute: Paolo and Laura Mora, to name just the two best remembered. Giorgio

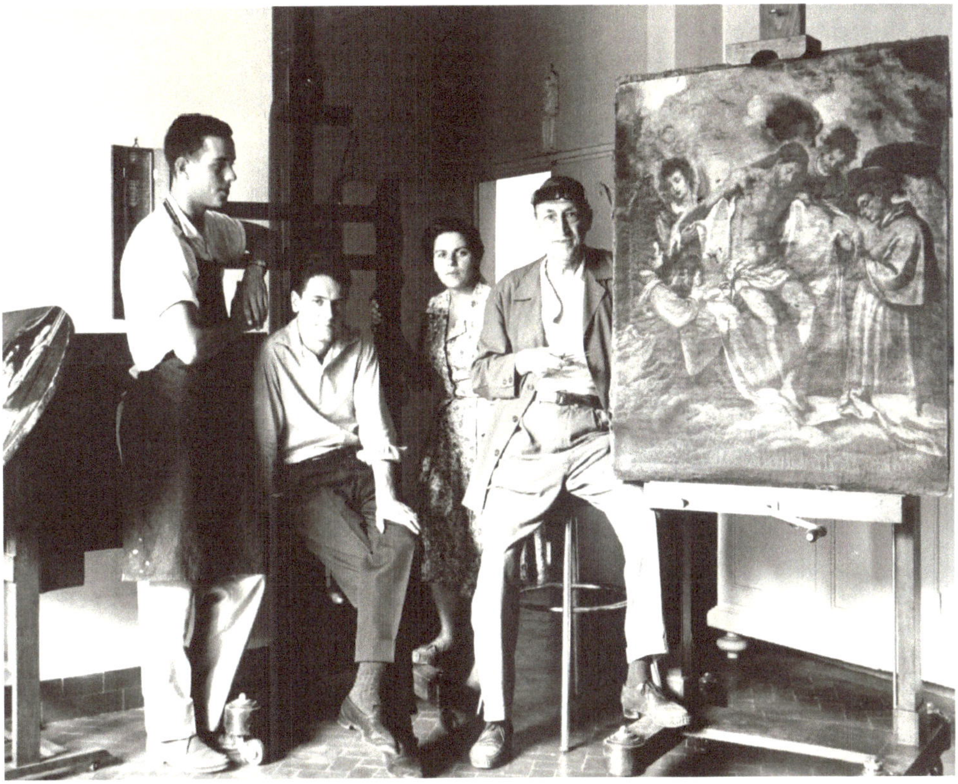

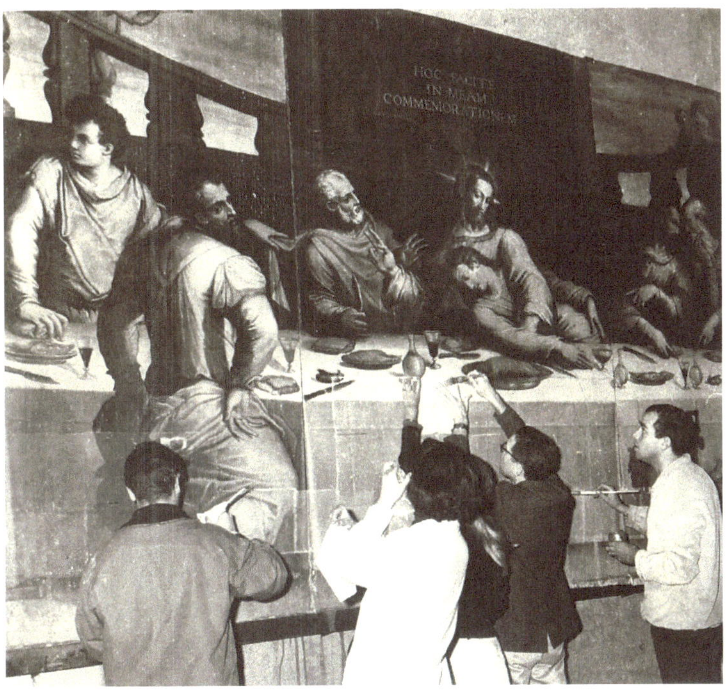

Firenze, 18 giugno 1958

Egregio Prof. Brandi,

per una assenza prolungata più del previsto ho avuta soltanto ora notizia della Sua richiesta sui fissativi usati nella Cappella Scrovegni, e provvedo subito scusandomi del ritardo.

8 - Per i colori duri come il bianco, il verde, ecc., che si sollevavano in piccole scaglie ho usato di preferenza il POLISTIROLO. Questa resina che è una delle meno raffinate presenta caratteristiche tali da farla preferire spesso ad altre più fluide. Proprio per la poca penetrabilità di essa si riesce con fatica a farla insinuare in crepe già aperte, ma siamo certi che non si introduce a nostro dispetto in colori che non hanno bisogno di essere rafforzati e lascia perciò a questi tutti il respiro che è indispensabile a tutte le pitture murali.

Il difetto più grave del polistirolo è di ingiallire molto se rimane in superficie, perciò deve essere curata molto la pulitura dopo il consolidamento: cosa del resto molto utile anche quando si tratti di altre resine sia naturali che sintetiche. Per quanto mi risulta non perde mai la reversibilità.

In altri colori più polverulenti come gli azzurri, rossi, ecc. dove sarebbe stato impossibile rimuovere i residui rimasti in superficie ho usato la ISOBUTILMETACRILATO molto diluita e applicata con un sistema che ne fa affiorare il meno possibile. Poiché questi colori non si possono tormentare con lavaggi neppure dopo che sono fissati, si deve aver cura che il fissativo venga assorbito nell'atto stesso dell'applicazione.

- Per gli intonaci e le croste più dure, ho usata con risultati ottimi la emulsione vinilica VINAVIL, che per la presenza dell'acqua da essa contenuta rendeva momentaneamente plastiche le zone da trattare e permetteva di riportarle più vicine alla sede originaria senza frantumarle. — 7

Ora, a distanza di un anno, avrei molto piacere di poter controllare anche con Lei i risultati di questa operazione. Spero perciò in un'occasione che ci permetta di realizzare questa indispensabile collaborazione.

Molti cordiali saluti

Leonetto Tintori

Istituto Centrale del Restauro
ARCHIVIO
Pos. 2.4.1
N. di Prot. 1206
Data 10-7-1958

Torraca, an accomplished visiting chemical engineer, lent invaluable technical expertise to most undertakings. Moreover, these projects were invariably guided by the thoughtful historical perspective and unerring taste of Giovanni Urbani who was to succeed Brandi as director in 1961.

What seemed particularly urgent to a young student at the ICR in those years was the need to explore the qualities and possible applications of a plethora of new synthetic materials issuing from the laboratories of large industrial firms such as Dow, DuPont and BASF. It was exciting to imagine that, on the heels of the perceived advances achieved in Germany and the United States with the use of natural and microcrystalline waxes, evolving categories of man-made resins would provide practicality and stability as varnishes and consolidants.

I was fascinated to discover among my papers a number of documents dating from those years. They vividly illustrate what can only be described as the rapt and intense scrutiny being devoted to an ever-expanding range of polymers. Some random notes – and there are many more pages – confirm the vast variety of products being investigated and tested. Particularly significant is a very early reference to Paraloid B-72, a resin evidently first suggested by Dr Robert Feller of the Carnegie Institute, with Tintori again acting as the 'Florentine liaison' to his Roman colleagues.

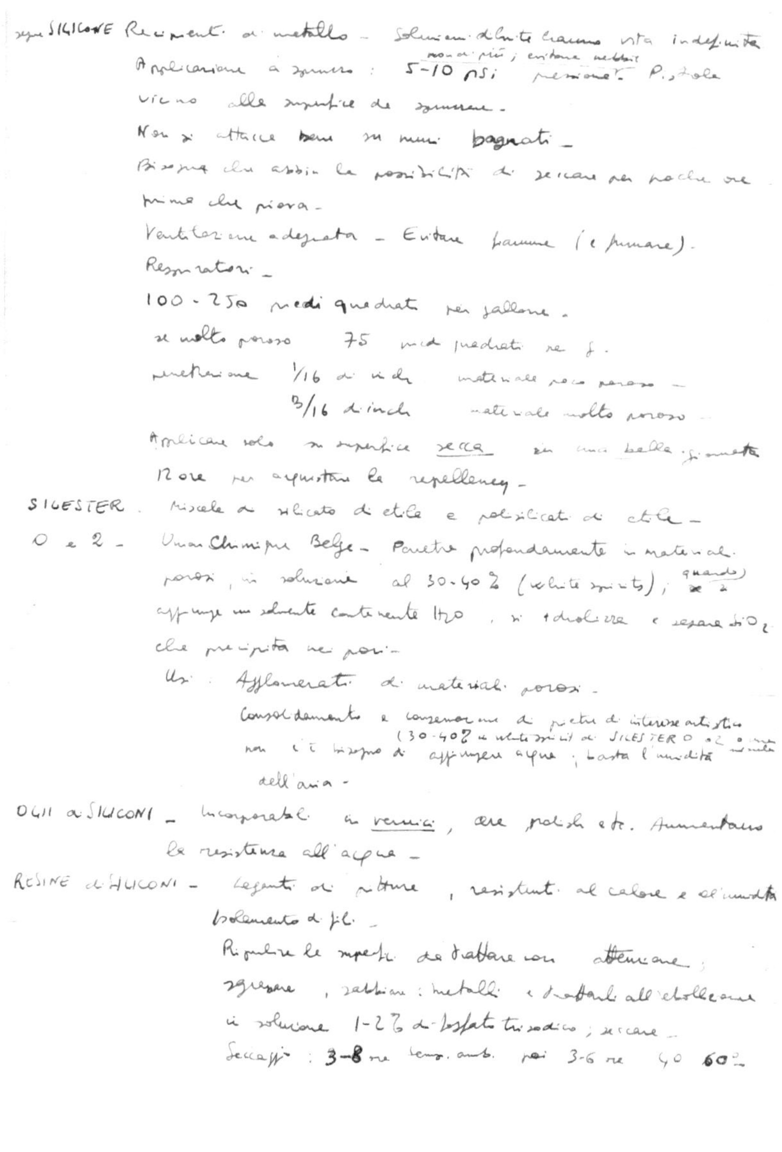
sopra SILICONE Recipienti di metallo – Soluzioni diluite hanno vita indefinita
non di più; evitare nebbie
Applicazione a spruzzo: 5-10 psi pressione. Pistola
vicino alle superficie da spruzzare.
Non si attacca bene su muri bagnati –
Bisogna che abbia la possibilità di seccare per poche ore
prima che piova.
Ventilazione adeguata – Evitare fiamme (e fumare).
Respiratori –
100-250 piedi quadrati per gallone.
se molto poroso 75 piedi quadrati per g.
penetrazione 1/16 di inch materiale poco poroso –
3/16 di inch materiale molto poroso –
Applicare solo su superficie secca in una bella giornata
12 ore per acquistare la repellency –
SILESTER. Miscele di silicato di etile e polisilicati di etile –
O e 2 – Una Chimique Belge – Penetra profondamente in materiali
porosi, in soluzione al 30-40% (white spirits); quando si
aggiunge un solvente contenente H_2O, si idrolizza e separa SiO_2
che precipita nei pori –
Usi: Agglomerati di materiali porosi –
Consolidamento e conservazione di pietre di interesse artistico
(30-40% in white spirit di SILESTER O o 2 o una miscela)
non c'è bisogno di aggiungere acqua, basta l'umidità
dell'aria –
OLII di SILICONI – Incorporabili in vernici, cere polish etc. Aumentano
la resistenza all'acqua –
RESINE di SILICONI – Leganti di pitture, resistenti al calore e all'umidità
Isolamento di fili –
Ripulire le superfici da trattare con attenzione;
sgrassare, sabbiare: metalli e trattarli all'ebollizione
in soluzione 1-2% di fosfato trisodico; seccare –
Seccaggio: 3-8 ore temp. amb. poi 3-6 ore 40-60°

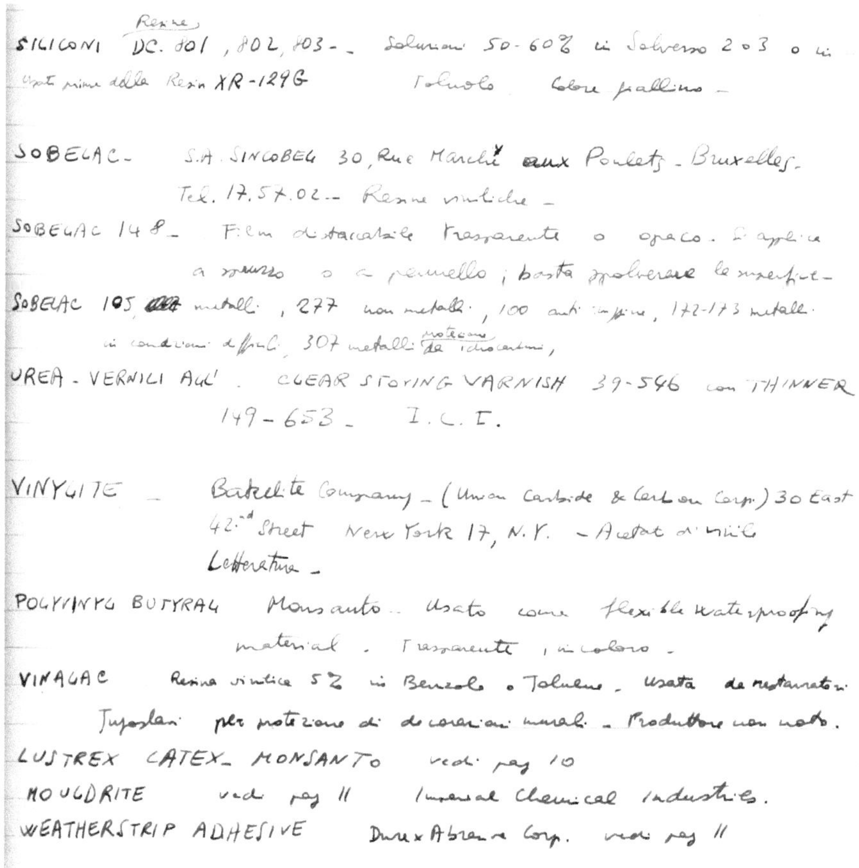
Resine
SILICONI DC. 801, 802, 803 -- Soluzioni 50-60% in Solvesso 203 o in
Usate prima della Resin XR-129G Toluolo Colore giallino -

SOBELAC - S.A. SINCOBEL 30, Rue Marché aux Poulets - Bruxelles.
Tel. 17.57.02 -- Resine viniliche -
SOBELAC 148 - Film distaccabile trasparente o opaco. Si applica a spruzzo o a pennello; basta spolverare le superfici -
SOBELAC 105, metalli, 277 non metalli, 100 anti ruggine, 172-173 metalli in condizioni difficili, 307 metalli protezione da idrocarburi,
UREA - VERNICI ALL'. CLEAR STOVING VARNISH 39-546 con THINNER 149-653 - I.C.I.

VINYLITE - Bakelite Company - (Union Carbide & Carbon Corp.) 30 East 42nd Street New York 17, N.Y. - Acetat di vinile
Letteratura -
POLYVINYL BUTYRAL Monsanto.. Usato come flexible waterproofing material. Trasparente, incoloro -
VINALAC Resina vinilica 5% in Benzolo o Toluene - Usata da restauratori Jugoslavi per protezione di decorazioni murali - Produttore non noto.
LUSTREX LATEX - MONSANTO vedi pag 10
MOULDRITE vedi pag 11 Imperial Chemical Industries.
WEATHERSTRIP ADHESIVE Durex Abrasive Corp. vedi pag 11

Firenze 11 maggio 1960

- Discussione con il Prof.Tintori. Conclusioni.-

Ognuno dei fissativi présí in considerazione presenta notevoli difetti: dopo una cernita preliminare le possibilità attuali sono state ridotte a tre:

1) Metacrilati in soluzione
2) Calathon AC
3) Metacrilati in emulsione

1) Metacrilati in soluzione.-

Due prodotti sono possibili:

- Lucite 45 duPont de Nemours (metacrilato di isobutile)
- Paraloid B 72 Röhm & Haas (metacrilato di etile e metile)

Per la Lucite resta il dubbio della insolubilizzazione nel tempo. (Ricerche della National Gallery).-

Il Paraloid dovrebbe essere provato per vedere se non contenga dei plastificanti, che potrebbero essere dannosi. Il Prof.Feller ha dato assicurazioni sulla permanenza della solubilità del Para= loid dopo lunghe prove di esposizione. L'effetto estetico del Paraloid secondo il Prof.Tintori è meno buono, ma esso può essere ridotto al minimo asportando qualsiasi residuo di fissativo dalla superfice dopo il fissaggio (precauzione che è comunque necessa= ria per evitare la formazione di pellicole che potrebbero provoca re lo strappo del colore).-

2) Calathon AC.-

Non è molto stabile nel solvente usato (acqua : alcool 70 : 30) e tende a coagulare durante l'impiego forman= do delle pellicole opache che possono dare un cattivo effetto

= 2 =

estetico sotto alcuni angoli visuali.-
I dati sul suo invecchiamento sono ancora incompleti, co= munque non confrontabili con quelli sui metacrilati, che or= mai sono stati provati per anni dall'industria e sono consi= derati le materie plastiche più stabili.-

3) Metacrilati di emulsione.-

Due possibilità:

- Primal AC 55 Röhm & Haas
- Emulsione preparata da noi

L'uso del Primal è interessante: la Röhm & Haas ha dati di esposi- zione di 5 anni, molto soddisfacenti. Richiede però l'aggiunta di un antimuffa.

Occorrono però delle prove per stabilire se gli emulsionanti che il Primal contiene possono essere dannosi.

Sarebbe importante preparare emulsioni con emulsionanti sicuri (ad es. il Carbovax 4000 o 6000), queste emulsioni sarebbero meno stabili di quelle commerciali, ma potrebbero facilmente essere preparate di volta in volta. Occorrono però esperimenti e questa soluzione può essere presa in considerazione solo per il futuro.

L'Istituto dovrà effettuare al più presto una prova nel Weather-O= meter con:

Paraloid B	72
Lucite	45
Calathon	AC
Primal	AC 55

Per il Calathon AC occorrono delle prove ulteriori per un cam= biamento di solvente che ne faciliti l'impiego da parte del restaura tore (va tenuto presente che con un altro solvente l'effetto este= tico potrebbe cambiare).

Se occorresse una decisione immediata si dovrebbe preferire il Paraloid B 72, riservandosi di cambiare fissativo se gli esperimen ti in corso indicassero che un altro prodotto ha caratteristiche migliori.-

SOME QUESTIONS FOR MR FELLER

1) Mr. Tintori showed us in Florence a letter by Mr Feller including some information on the ageing of poly methacrylates. Ethyl- and Methyl- polymethacrylates do not appear to undergo cross-linking to a large extent under the action of ultra-violet radiation.

We assume that Propyl- , Isopropyl-, Butyl- and Isobutyl- polymethacrylates will udergo cross-linking after ageing, or at least there is a danger that they will.

Is our assumption correct?

2) It is possible to perform some 'reforming' operation on cross-linked Isobutyl-poly methacrylate, so as to induce some solubility (or at least swelling) in the polymer film?

(We are investigating the possible use of acrylic polymers as fixatives in the frescoes of the Scrovegni Chapel in Padua; the point is that the fixative should be removable with relative ease after a number of years or even possibly decades).

3) Does Mr. Feller know what kind of polymethacrylate is used in the formation of Rohm and Haas' Paraloid B-72? Is there available any data about its solubility behaviour after ageing?

By the time the flood struck Florence, many of these materials had begun to find wide application in studio practice in Italy. The very fact that they represented a new and more up-to-date 'technological' approach to traditional conservation methods held particular appeal at a time when Italian society as a whole was experiencing a headlong rush towards 'modernisation'. Paraloid was to play a particularly significant role in the emergency measures taken immediately following the water's retreat (actually a thick, emulsified amalgam of water, heating oil, organic and inorganic detritus).

Since a great number of panels had been either partially or wholly submerged for upwards of five or six hours, it was obvious that the massive losses of colour resulting from this first dreadful trauma would be far exceeded if the wooden supports were to be allowed to dry without drastic preventive measures. I attended one of several emergency meetings held at the Uffizi and open to all who had some measure of experience and happened to be available. Not surprisingly, the atmosphere was positively feverish; the sense of foreboding was magnified by the evident inability, on all sides, to provide any concrete parameters on which to base critical decisions. Quite simply, no one was in a position, on the basis of personal experience, to describe or even imagine a recommended treatment.

In this atmosphere of near panic two basic decisions were taken: first, to 'paper' the painted surfaces of all affected panels (as well as many canvases) with Paraloid B-72 in a toluene solution; secondly, to place all affected works as rapidly as possible in environments of controlled relative humidity. Ever since those fateful days, legitimate questions have been asked: were these the *only* solutions possible and with the *only* materials possible? Had I been asked – which, of course, I was not – I would undoubtedly have answered in the affirmative; after all, I had been as fascinated as my superiors with the potential promise of the new materials I had studied. The rest of the story – the long litany of problems stemming from those early decisions – has yet to be written, and the

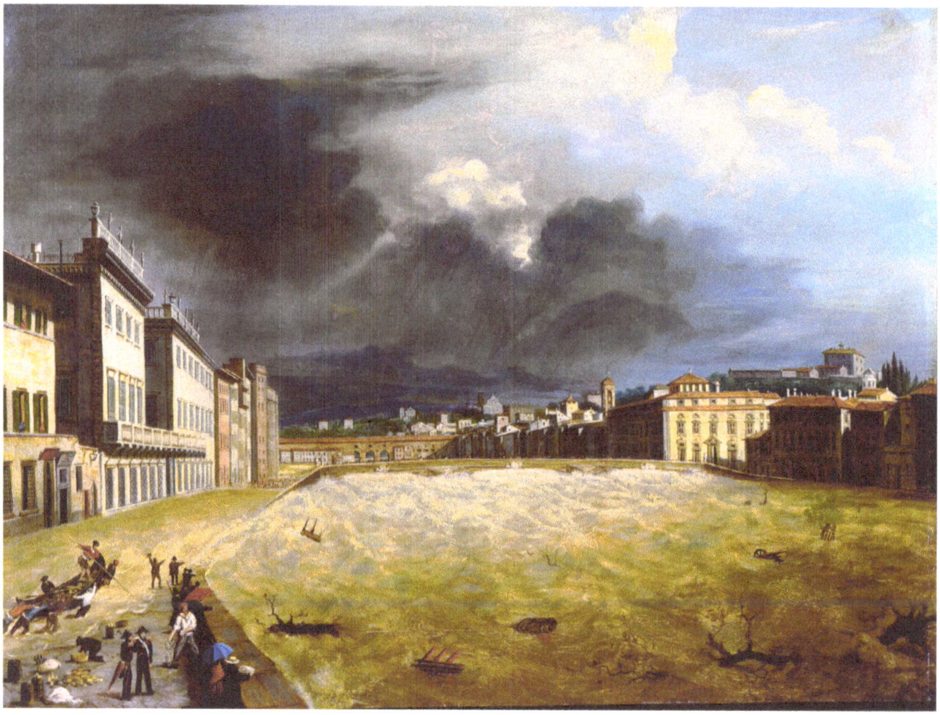

The Flood of November 1844 (Grassi Collection, NY)

conclusions yet to be drawn. In this context, I was particularly struck by the coincidence that it was in this 40th anniversary of the flood that the great Vasari ***Last Supper***, which I had helped 'paper' all those decades ago, was finally reinstalled!

The anniversary was also an occasion to reflect on what the future might hold; after all, even though it was by far the most destructive, the 1966 flood was hardly the first of such calamities to befall the city (see illustration above). Can we hope to have learned something from those events 40 years ago? Unfortunately, some contributors at the symposium were not so reassuring.

Restoration in Florence following the flood

Giorgio Bonsanti

It cannot be denied that great advances were made in restoration techniques in Florence and Italy after the flood. In many cases these were not simply chronological developments but were also the direct result of specific situations. The flood stimulated research into new methods of restoration; it also served as a catalyst, creating the ideal conditions in which to put into practice and accelerate certain procedures and reactions although, naturally, not all the innovations in Italian restoration at this time were directly related to problems posed by the flood.

At the time of the flood, restoration work in Italy – in Florence in particular – had reached the highest levels of excellence, partly because restorers in Italy, more than in any other country, had the advantage of centuries of experience and were frequently engaged in highly demanding restoration projects. It is also true to say that exchanges with other countries were fairly limited. The flood changed all that and Florence rapidly became de-provincialised, something for which, despite the unfortunate circumstances, I am eternally grateful. Florentines traditionally excelled in the restoration of paintings on panel and on canvas. As a result of contributions made by the great restorer Augusto Vermehren, they had also begun, as early as the 1930s, to experiment with radiography, a technique already well advanced in the UK, Germany and the United States. The Gabinetto di Restauro della Soprintendenza, the restoration department in Florence, was not founded until 1932, later than similar institutions in other countries. In Italy there was also a strong tradition of the restoration of mural paintings, and in these two sectors there were restorers of the highest calibre: Gaetano 'Nino' Lo Vullo and Dino Dini to name but two.

In other areas of restoration, it has to be admitted, methods in Italy lagged behind, even though Italian restorers were always distinctive for their individual techniques. In the restoration of stone for example, and of similar materials such as terracotta, we were able to perform complex and difficult operations, and the Opificio delle Pietre Dure was summoned to offer assistance throughout Italy, from Naples to Bolzano. We did, however, lack knowledge of the use of chemistry applied to consolidation and protective treatments, applications always to be approached with caution, but which could neither be ignored nor undervalued. The same was true of our expertise in the treatment of bronze and other metals. And it was in these two areas, in stone and metal conservation, that we benefited enormously in the short and long term from the international contacts established at the time of the flood. Restorers worldwide showed extraordinary solidarity, generously sharing their expertise and responding fully to our urgent need for help. I remember in particular the exchanges with experts such as Kenneth Hempel of the Victoria and Albert Museum in London, and with specialists from the Waffensammlung in Vienna. This spirit of openness and exchange influenced the newly formed, modern Opificio, founded in 1975. There, the belief in a close relationship between science and restoration, along with the potential of the best work in Italy, and Florence specifically, were brought to bear at every stage of research and directed towards the proposed intervention.

I should like to consider an exemplary case of the focused application of research following the flood, which produced results that became a point of reference in international restoration. This case concerns Enzo Ferroni, a respected chemist and former rector of Florence University, and Dino Dini, an exceptionally fine restorer with whom I had the honour to work on a daily basis for a number of years, and who worked on the restoration of the frescoes by Fra Angelico in the Convent of San Marco – an operation completed in 1983. I do not know how far these two were aware of the research into the consolidation of porous surfaces with inorganic substances carried out in England since the nineteenth century, but never widely adopted. Certainly their decision to use inorganic materials in consolidation, specifically barium hydroxide, marked the beginning of this method becoming a feature of Florentine restoration of mural paintings, in contrast to methods adopted in Rome at the Istituto Centrale per il Restauro (ICR). Their choice was largely dictated by the conditions created by the flood, which made the use of acrylic resins extremely problematic.

These resins had been introduced in the early 1960s on the initiative of the famous restorer from Prato, Leonetto Tintori, who was in contact with the restoration centres at New York University (NYU) and Pittsburgh, and with the noted restoration chemist Robert Feller. Acrylic resins, the most well known being Paraloid (Acryloid in the United States), formula B-72, were then applied to such major works as the fresco cycle of Piero della Francesca in Arezzo with an alacrity which I still find alarming. Put simply, these resins were applied in much the same way as *beveroni*, the glues used for centuries to consolidate wall paintings with crumbling and flaky surfaces. These resins covered the painted surface in an adhesive film that stabilised the surface vertically by attaching it to its support, as well as horizontally, forming an unbroken protective layer. At the time of the flood, naturally, many of the deleterious long-term effects of this treatment were unknown: the strictly theoretical 'reversibility' became extremely difficult to achieve with the passage of time; humidity became trapped within the wall because the surface was no longer porous; and there were dramatic alterations to the colours. Paraloid was, however, used extensively after the flood in facing easel paintings with tissue paper where it created problems not related to the substance itself but to the length of time the facing remained in contact with the painted surface. This posed severe problems when it came to its removal.

But to return to mural paintings, it was impossible to apply acrylic resins to entire walls soaked in water in which there was an worrying increase in the quantity of salts. Some of these, such as nitrates, could not be eliminated and could only be partially neutralised by the application of tributyl phosphates. Nor was it immediately possible to adhere canvas to the painted surfaces in order to detach the frescoes from the wall. When restorers were able to do this they achieved some really remarkable results, unparalleled in the history of restoration and culminating in the detachment in one piece of the *Last Supper* by Taddeo Gaddi in the refectory of Santa Croce on 10 October 1967: 122 square metres of fresco without a break in the surface. This was an unprecedented operation and will probably never be repeated – the present policy is to consolidate *in situ* wherever possible and to avoid detachment. But at that time, between 1966 and 1968, more than 3000 square metres of fresco were removed from their supporting walls where had they remained they would have been lost forever.

In the fight against time and in those conditions, detachment was frequently the only choice available. As an alternative however, as I mentioned earlier, experimentation with the use of barium hydroxide for the consolidation of wall paintings increased and was applied extensively for the first time to the great *Crucifixion* by Fra Angelico in the chapter of the convent of San Marco. This brilliant solution exceeded all expectations and eliminated the sulphates that had reduced the calcium carbonate in the original plaster to powdered chalk. This resulted in a restored level of carbonate in the wall, even though initially (later the chemical transformation could advance further) it was barium carbonate rather than calcium. At the pre-consolidation stage, in smaller concentrations, barium was used first on the Piero della Francesca cycle in Arezzo. This was of vital importance to allow cleaning to be carried out, especially in cases where the surfaces were in a particularly critical condition and requiring urgent consolidation. All these techniques, conducted alongside work in the UK on the possibility of recreating carbonate layers, identified and developed 'soft' treatments for wall surfaces, and also in some instances for stone. This work was carried out with the underlying conviction of the paramount importance of the theoretical principle of the compatibility of materials, and in adherence to the concept of minimum intervention – two vital considerations that space prohibits me from discussing further in this paper.

In order to clean and rid the stone of the polluting substances left by the flood, experimental packs of absorbing clay, such as sepiolite and attapulgite, were applied. It is worth remembering that in the 1970s, enormous progress had been made in Italy concerning the treatment of stone, and this of course made its impact on Florence. I cannot here describe the advances in archaeological, book or archive conservation as it is beyond my specific field.

The general attitude of profound respect for the intrinsic properties of each work gradually permeated research into the restoration of all kinds of materials. The notorious cradles on panel paintings have been replaced by far less invasive supports, developed in

Italy, most noticeably in the Opificio workshops. The centuries-old practice of blocking the natural movement of the wood was abandoned in favour of installing perpendicular supports, both effectively reinforcing the structure and preserving its natural elasticity. A particularly remarkable example of this technique is provided by the support given to Giorgione's *Castelfranco* altarpiece, restored a few years ago by Pierpaolo Monfardini of Bologna. Furthermore, two large panels damaged by the flood, by Bronzino and Francesco Salviati, are about to be reinstalled in the Museo di Santa Croce with fully intact supports, the restorers having avoided one of the typical operations adopted in the past: the *trasporto* or removal of the pictorial layer from the support with irreparable damage to the latter. Their achievement, absolutely unprecedented in the case of paintings that were covered up to two-thirds of their height in water, was possible through a global approach to a specific problem, always considering the work in the context of its immediate environment. These interventions on the panel are complemented by an ingenious but simple technique: the creation of a box around the sides and back of the panel in order to improve insulation. This was first tested on a sixteenth-century Ferrarese painting in the Capodimonte Museum in Naples, the *Deposition* by Ortolano, and then on the *Coronation of the Virgin* by Botticelli in the Uffizi, a painting unscathed by the flood but which had long suffered from a dramatic lack of adhesion between the various structural layers.

The large lacunae left in paint layers when the floodwater receded forced all those involved in conservation to reconsider the theories, techniques and methods of performing inpainting. The late Umberto Baldini therefore devised proposals for the restoration of paintings based on what he defined as 'selection' and 'chromatic abstraction'. I should like to mention at this point that Baldini's methods should not be considered an alternative to, but rather the development of, the *tratteggio* technique, known as *rigatino*, recommended in about 1950 by the ICR under Cesare Brandi. In Florence there is a fine example of restoration using this technique by Laura Mora of the ICR carried out in the early 1950s and which I still consider exemplary: Fra Angelico's *Deposition* in the Museo di San Marco. In my opinion, chromatic selection produced results of the highest quality, even in particularly challenging works such as the magnificent *Coronation of the Virgin* by Lorenzo Monaco in the Uffizi.

Finally, I should like to consider the treatment of the *Gates of Paradise* by Lorenzo Ghiberti. The flood damaged the structure of the doors from which a number of the panels were knocked out of position. An overall assessment was essential so that all the problems related to the restoration could be carefully considered. Apart from anything else, this allowed for a detailed understanding of Ghiberti's technique in casting and fitting the individual panels, including the smaller ones inserted into the door frame. This research eventually led to the development of new laser technology for cleaning gilt bronze, which also allowed the removal of unwanted corrosion products. Experimentation with lasers began in Florence in about 1990, and the city has become the most advanced centre in Italy for the use of this technology in restoration. The restoration of the *Gates of Paradise* was one of the most demanding projects undertaken in the last century. It presented an extraordinary variety of problems but also gave birth to diverse solutions. For that reason it can be considered emblematic of the commitment shown and consistent quality achieved in Florentine restoration in the 40 years since the flood.

Editor's note

Translation by Heather Roberts.

The rescue of statues and sculptures in Florence

Kenneth Hempel

In the wake of the flood, the British response was immediate and I was just one of the many British experts who were sent to Florence as part of the joint forces of the British government of the day and the newly formed Italian Arts and Archives Rescue Fund, under the chairmanship of Sir Ashley Clarke, former British ambassador in Rome.

The Victoria and Albert Museum's former Keeper of Architecture and Sculpture, and later director of the V&A, Sir John Pope-Hennessy, was asked by the Rescue Committee to go to Florence and assess the damage caused by the flood to the city's great sculptures and monuments. Sir John, in his autobiography ***Learning to Look***, gave a detailed and accurate account of the widespread damage that had been inflicted. Incidentally, Sir John also described Villa La Pietra in his book: 'The huge villa was a cool oasis.' He was a very good friend of Sir Harold Acton and a regular guest at La Pietra.

In 1966 at the time when the flood occurred I was the restorer in charge of the conservation of the Italian Sculpture Collection at the V&A. When Sir John returned from Florence he was very saddened by all the monuments and sculptures that were in desperate need of restoration. He summarised the facts and the events that resulted in my involvement and that of the V&A in the rescue of the Florentine sculptures as follows:

> But there were two areas in which the salvage operations were notably deficient; one was the restoration of sculpture and the other that of the applied arts. The Conservation Department of the Victoria and Albert was in a state of high efficiency at the time. Its head, Norman Brommelle, had been trained as a painting restorer at the National Gallery and had reorganized the entire Conservation Department, and was married to Joyce Plesters, the head of the Research Laboratory at the National Gallery. I had, moreover, for some years been working with a specialist in sculpture restoration, Kenneth Hempel, who, when the flood occurred, was the most accomplished restorer of Italian sculpture in the world.

> The Conservation Department naturally had much experience in the purchase of equipment and materials (experience of which the Soprintendenza in Florence seemed to be woefully lacking) and a first-rate administrator named Gorton, who was himself an experienced restorer of furniture and woodwork. Help in these areas could be given to the Florentine authorities, provided that the Ministry of Education allowed members of the museum to work in Italy without formality and provided that funds, for travel, subsistence, and the purchase of materials, were made available by the London Committee. When I returned to London both conditions were met. The Italian Art and Archives Rescue Fund, under the vigorous chairmanship of Ashley Clarke, gave the project full support, and the attitude of the Establishment Division of the Ministry of Education was also cooperative and liberal. In Florence, I explained, with the Bargello in mind, that when disasters occur, it is best to accept them as portents, and to adopt the target, not of restoring the status quo, but of improving on it. Hempel therefore inaugurated a broadly based campaign for cleaning marble sculpture.
>
> The results achieved were impressive and positive. With the invaluable help of Kirsten Aschengren Piacenti, a restoration studio was set up in the Palazzo Davanzati, and it was there that Hempel and Joyce Plesters jointly discovered the first traces of gilding in the hair of Donatello's wooden Magdalen.
>
> It is difficult today to visualize the worldwide sense of sympathy and compassion that the flood aroused. Numbers of students found their way to Florence anxious to help. As one looks back over the lists of contributions, especially small contributions from the student bodies of colleges and schools of art, the enterprise reads like an affirmation of faith (J. Pope-Hennessy, *Learning to Look*, New York, Doubleday, 1991, pp. 157–8).

I had built up an intimate knowledge of the nature and range of materials of the Italian and, most especially, the Tuscan statues and sculptures during many years preceding the 1966 flood. I had a good background on the most appropriate treatments that could be used to restore marble, stone, terracotta, gesso, stucco and wax-finished statues and sculptures (plain, gilded or painted) by using the most suitable conservation materials, chemicals and techniques. This gave me a sound experience of the options and the flexible approach that would be needed to adjust my restoration and conservation methods to the circumstances and to the scale of intervention, depending on the degree of the state and damage suffered by the sculptures.

The experience I gained at the V&A evolved into the scientific conservation methods based on the various treatments that I was able to carry out such as cleaning, consolidation, bonding, dowling, filling, modelling, retouching and protection. These were achieved with conservation materials that had been tested in the V&A laboratory and other museums, a summary of which is given below. Due to these scientific methods and techniques that I had developed and used to restore and conserve Italian statues and sculptures as well as small monuments in the V&A Collection, I was in a strong position to help with the problems that Florence was suffering following the flood.

As a civil servant I required official permission from the British government before I could go to Florence to offer my help and expertise. This took a certain amount of time. While I was waiting for permission to come through, I conducted an experiment at the V&A conservation laboratory in order to try to find the best method to extract the oil stains from the sculptures. I first attempted to reproduce a similar oily, muddy, polluted 'gunge' that had damaged the Florentine artistic heritage. But of course I was limited to new marble samples. Sir Harold Acton in his memoirs gave a good description: 'The invasion of oily mud ... the deposit of thick black oil above doors and windows ... the sewers vomiting their contents ... the battered and overturned cars, the huge trees and oil drums.' Obviously I could not reproduce the exact 'mixture' but I came close.

In a large waterproof tank half filled with water I added a bucket full of soil then half a bucket of crude oil, horrible thick black stuff and human urine, mixing them thoroughly to simulate the contaminated floodwater of the Arno. Into this revolting concoction I placed several core drillings of new Carrara marble, 6–7 cm in diameter and 80 cm long (I could not, of course, use antique marble). I then left the marble pieces to soak up the black mixture in the tank for several days until they became black with oily mud stains.

One word about marble for those not closely intimate with this material: marble starts life as a limestone stratum deep beneath the earth's crust. Then it undergoes a massive change by being heated to a liquid form due to volcanic activity and as it cools down it forms calcium carbonate crystals. The size of the crystals depends on how quickly or slowly the mass cools down: slow cooling produces large crystals such as Greek marble while small crystals such as Carrara marble indicate a faster cooling. Contrary to popular belief, marble is a soft stone measuring only 3 on the Mohs scale. Also, it is absorbent, allowing water, oil and other liquids to penetrate as much as an inch in a matter of only 12 hours.

To clean the marble samples stained by the oily mud of the experimental tank, I had to use a solvent to dissolve the crude oil. The sepiolite was mixed first with white spirit, then with Shellsol A, and so on, using different solvents. I then coated each oily, stained marble sample separately with a mixture of sepiolite about 1 cm thick. The creamy poultice clings to the marble even in an upright position – very useful for use on vertical surfaces.

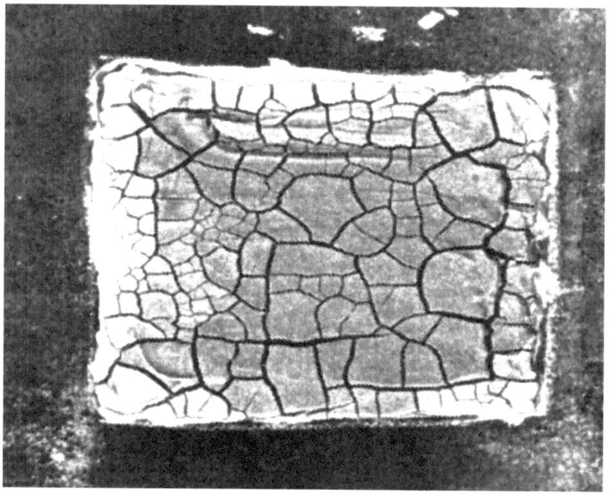

The mudpack dried out, forming a craquelure as sepiolite resembles clay. This was removed and any residue left on the marble surface was washed off with some solvent used in the mix. The experiment worked. The best results were given by the mixture of sepiolite and Shellsol A – the marble samples had returned to their original colour and appearance.

Chemically, sepiolite is a trisilicate of magnesium, and in powder form of 100 mesh it is a very absorbent substance indeed. Seen under the microscope, the powder consists of tiny needle-like crystals which have the capacity to hold its own volume of liquid. In other words, you can pour a bucket of liquid into a full bucket of sepiolite without any increase of volume. No spillage and no overflow will occur. A brilliant V&A chemist introduced me to this material many years before the flood. In those days, sepiolite was used in England in large quantities on the shop floors of big factories to suck up and absorb the lubricant oil from heavy industrial machines. As the experiment gave good results, I was confident that the treatment would be effective in Florence.

With all the permissions and all the arrangements finalised, I was able to leave London at last, though on my arrival in Florence I realised the enormity of the difficulties that I had to overcome which were very testing. I had the help of many students from all over the world especially from Italy, England and America in this monumental task. Also, of course, the V&A staff, especially the Architecture and Sculpture Department and the Conservation Department, were of great help during this time.

I arrived in the depth of winter to find that in Florence conditions were still very harsh. I was met by Dottoressa Kirsten Aschengreen Piacenti, also known as Cristina, who to my great relief spoke English. Unfortunately I did not speak Italian, and after all these years still do not speak the language. She took me first to the hotel where I deposited my personal luggage. Then Cristina accompanied me to Palazzo Davanzati where the large box containing my conservation materials and tools had just been delivered. I was shown a rather large, dark, cold and entirely empty room on the third floor of the *palazzo*.

I began by asking Dottoressa Piacenti 'Where are the restorers?' 'We do not have marble-stone restorers', she replied. I had been warned of this in London but I did not believe it! In Italy, in Florence, there were no stone sculpture restorers? This really hit me hard. I learned later that monument and sculpture restoration was, in fact, performed by removing the decayed pieces of marble or stone and replacing them with new marble or stone carved to a very high degree. Of course this was *not* conservation.

It was then agreed with the superintendent, Professor Ugo Procacci, that I would first carry out a test to see if my cleaning methods would work in Florence. The room at Palazzo Davanzati was not ideal: it was poorly lit, had no running water and there was no direct electricity supply and no heating. I had to wear my sheepskin coat and hat even indoors because of the intense cold. Eventually a kind custodian provided a one-bar electric fire on a long extension lead connected to another room of the *palazzo*, which really did not make much difference.

Then a group of workmen arrived heavily laden, carrying up the stairs two flood-damaged tombstones, which were severely blackened. I had to improvise a sort of a bench from some odd wooden boxes as best I could and then set to work. I mixed the sepiolite and Shellsol A to a thick cream and applied it to the tombstones. While I waited for the mudpack to dry I left Palazzo Davanzati to return to the warmth of my hotel for lunch. I returned in the afternoon to check if the poultice had dried out, but it seemed that the cold temperature had slowed the drying process of the solvent so I left again.

By the following morning the telltale 'craquelure' had appeared on the mudpack. It was ready to be removed. As I did so, I saw that the tombstones were still black. I had to repeat the application of sepiolite mixed with the solvent many times before the black oily stain eventually started to 'budge'. My persistence paid off. I was relieved to have succeeded and celebrated with a good strong cup of coffee at the 'bar' across the road. (Otherwise I would have had to return to London with my tail between my legs.)

Cristina Piacenti was delighted with the result and we set out to organise a large-scale cleaning programme. The room on the third floor of Palazzo Davanzati was covered with linoleum. Trestles and tables were produced, and a stream of students appeared: some Italian, some English from the Courtauld Institute of Art, several from the USA and one from Mexico. It was as though a small Tower of Babel had descended on me! An English student fluent in Italian came to my aid, which helped me greatly in communicating with the Italian student workforce and also the officials.

More improvements and facilities were made available in the room with American and British funds: a sink with running water, improved lighting, an electricity supply and heating – outside in the road, puddles were still icy. The winter months were slow to pass; we had to order more conservation materials, mostly sepiolite and Shellsol A as the oil stains were everywhere. In our small studio we could only treat relatively small pieces of sculpture because we were on the third floor, but as spring approached I divided the students into small working groups and allocated them to numerous large projects all around the city centre.

The outdoor monuments and large sculptures exhibited different problems as for centuries they had been exposed to the effects of weathering and air pollution. The crystals in the old marble had loosened making the marble even more porous, therefore more susceptible to the ingress of dirt, water and gases. Oil, in particular, had penetrated deep into the structure of the marble where it remained for months. Removing it was not an easy task!

The floodwater had left a high tide mark of black oil on the marble sculptures. As the water retreated, it left oily marks on other absorbent parts of the marble. In some cases it was most peculiar to see the bottom half of a marble sculpture, bust or tombstone more or less clean, or, in some cases, coated with black crude oil whilst the top half exhibited a different type of dirt – blackened by centuries of accumulated grime and relatively modern air pollution crusts formed by sulphation. These different surfaces therefore required different cleaning treatments.

The flood had also damaged small models in terracotta, gesso, as well as stucco and wax. The force of the floodwater had smashed glass cabinets resulting in heaps of fragments all covered in black oily mud: a horrible mess. The demand for my attention and expertise on all these restorations and conservation projects grew to a point where I could not cope without extra help from the V&A. The head of the Conservation Department, Norman Brommelle, agreed to send reinforcements to help. Edwin Gorton and his assistant arrived with my assistant, Judy Larney. I was then able to dedicate more time and engage in more demanding and difficult restorations that the students were unable to carry out.

Certain of my restoration and conservation works as well as some of those that Miss Larney carried out under my direction (with the help of some of the students) are recorded in the catalogue of the exhibition of the restoration of sculptures and minor art objects held at the National Museum of the Bargello from December 1967 to February 1968. In this catalogue are statements by Professor Ugo Procacci, the superintendent, who in the preface refers to the restoration problems that appeared difficult if not audacious or even desperate. The catalogue offers a rigorous scientific documentation of achieved solutions helpful for future and similar works. The sculptures varied from the largest – by Bandinelli, Ammannati, Giambologna etc. – to the smallest models by Tribolo and Francavilla.

The restored 'rescued' objects were significant not only for their artistic value but also as examples of a more varied technical application of restoration methods performed by international collaboration in the new Palazzo Davanzati laboratory.

Professor Berti, director of the National Museum of the Bargello (on pp. 1–4 of the catalogue) gives accurate insights into the problems with which I had to deal at the Bargello. Cristina Piacenti carefully explains how the restoration centre of the Palazzo Davanzati played an important role in the rescue work that took place there. Dottoressa Raffaella Rossi Manaresi finally gives a detailed account of the scientific treatments used on the outdoor sculptures of the Bargello's courtyard and the indoor statues (pp. 10–23). Some of the statues and sculptures are also shown in the plates (I–XI).

The recording in the catalogue of my interventions is, in some cases, not very accurate as far as the use of certain conservation materials and chemicals is concerned (see treatment steps and materials listed on p. 12). Nevertheless it provides a written record of some of the work that I carried out in Florence using a flexible approach and adapting my conservation techniques and methods to suit the degree of damage suffered by each sculpture, adjusting the procedures to the specific problems I encountered.

The Palazzo Davanzati laboratory arose from a need for a restoration centre to treat relatively small sculptures. It was also used as the headquarters for coordination and preparation procedures for the restoration and conservation of large statues and sculptures *in situ*, whether indoors or outdoors. I introduced new concepts and conservation treatments never used in Florence before, which were acknowledged by the Florentine authorities.[1]

The 1966 flood in Venice and subsequent restoration projects

The great flood of 1966 was caused by exceptionally bad weather in the north of Italy. While helping in Florence, I was asked to visit Venice at the same time to witness the damage that the flood had caused in that city.

It is important to distinguish the differences between the type of flood experienced by Florence and that of the Venetian ***acqua alta***. In Florence, the Arno, fed by all its tributaries, broke its banks and defences, flooding the city violently and turning it into a turbulent fast-running riverbed. Venice was flooded in a different way: the sea level of the lagoon rose dramatically and it was simultaneously engulfed by the influx of a very high tide, strengthened by strong winds, and the discharge into the lagoon of four raging rivers, carrying the torrential rain that for many hours battered the hills and mountains around Venice.

The floodwater contamination was of different strength and quality: in Florence it was muddy, oily and fast running while in Venice the mud and oil was somewhat dispersed because there was more water – the sea water was heavily mixed with that of the rivers.

It must also be remembered that both Florence and Venice had suffered floods before. The incidence of severe floods in Florence appears to have been every 100 years, while in Venice they are a more seasonal event. Since 1966, they seem to recur more frequently at the exceptional high tides.

Careful consideration has to be given to the fact that these two cities were built with different materials and it is relevant to take into account the different structural, architectural and decorative way in which they were constructed. Venice was built on stilts in the soft marsh of the lagoon. Non-absorbent stones were used for foundations, columns and colonnades and also high plinths. Statues and sculptures, as well as marble and stone decorations, were placed well above ground level – above windows and doorways where floodwater would not reach – but the insidious sea water slowly contaminated and caused damage through saline rising damp, and the salty and polluted atmosphere affected the statues and sculptures equally. These factors made restoration and conservation in Venice a far more difficult and complicated rescue.

Sir John Pope-Hennessy also visited Venice in March 1967 on behalf of the British Fund and in his autobiography he records, 'in churches and museums it was difficult to distinguish damage due to negligence from damage attributable to the flood. Because of this it was decided to carry out in Venice a restoration project that presented every possible problem with the aim to raise the conservation sights. The church of the Madonna dell'Orto was chosen and with the Italian Art and Archives Fund was restored with great success.'

The restoration of this church included the statue of St Christopher on the façade above the church's main doorway. The ***in-situ*** restoration of this sculpture took place in the summer of 1969. The project was made possible because I had just trained a student from Venice, Giulia Musumeci, recipient of a British Council scholarship, in marble and stone restoration and conservation in London at the V&A. Miss Musumeci undertook the restoration of St Christopher, in Istrian stone, under my direction and with the help of the V&A Conservation Department. Istrian stone is a dense calcareous sedimentary limestone which is almost non-absorbent. However, it had been exposed to the elements and consequently had suffered from the extreme rigours of the Venetian environment and was affected by erosion and black sulphation.

The removal of the black sulphation crust was achieved by yet another technique – a method I had used before at the V&A – involving a very careful and sensitive use of an ultrasonic dental drill point (used by dentists to remove tartar). This technique was used after a preliminary light water spray wash.

The sculpture, at one and a half times life size, was at the maximum limit of what could be treated and cleaned in such a way, taking into consideration time and economical constraints, as the cleaning was done a few millimetres at a time and required extreme care and concentration, especially out-of-doors.

It became quite evident that new cleaning techniques and methods were needed for *in-situ* sculptures. This was achieved through more than 10 years of work and in cooperation with the Venetian authorities and with the aid of funds from Venice in Peril. It also required my full involvement along with that of the Conservation Department of the V&A and with the full collaboration and execution of the restoration by Giulia Musumeci, who is now my wife!

Our restoration and conservation of the Loggetta del Sansovino (also called the Loggetta del Campanile) was published in *Studies in Conservation* (vol. 22, no. 1, February 1977) and the restoration and conservation of the main entrance of the Venetian Ducal Palace, the Porta della Carta, was published in English and Italian by the Venice in Peril Fund.

Treatment steps and materials

The techniques employed in the conservation of Florentine statues and sculptures were based on the following treatments I used at the V&A.

Cleaning

- Sepiolite mixed with deionised water
- Sepiolite mixed with white spirit or Shellsol A
- Water-based Nitromors paint remover (based on methylene chloride)
- Hydrogen peroxide
- Ammonia
- Other solvents: Cellosolve (2 ethoxy ethanol)
- Acetone

Consolidation

- Polyvinyl alcohol and in some cases with polyvinyl acetate emulsion
- MS2A

Bonding

- Polyvinyl acetate emulsion adhesive
- Sintolit (a polyester resin)

Dowling

- Stainless steel dowels with Sintolit (polyester resin)

Filling and modelling

- Polyvinyl acetate emulsion DM2 and grains of alabaster

Retouching

- PVA – MS2A and dry pigments

Protection

- Cosmolloid wax mixed with white spirit or Shellsol A
- Finally a dusting with French chalk (talcum)

Examples of projects undertaken in Florence

A few examples of these materials and techniques used in Florence are given in the descriptions of the restoration and conservation of:

- The large marble sculpture group of *Florence Victorious over Pisa* by Giambologna
- The *pietra serena* relief of *Madonna and Bambino* by Jacopo Sansovino
- The terracotta model of *Honour Triumphant over Falsehood* by Vincenzo Danti
- The unbaked clay with wax finish model of *The River God* by Niccoló Tribolo

Restoration and conservation of the large marble sculpture group of Florence Victorious over Pisa *by Giambologna*

This group was flooded by the polluted oily black muddy water of the Arno up to *Florence's* chest. Two different cleaning techniques were required:

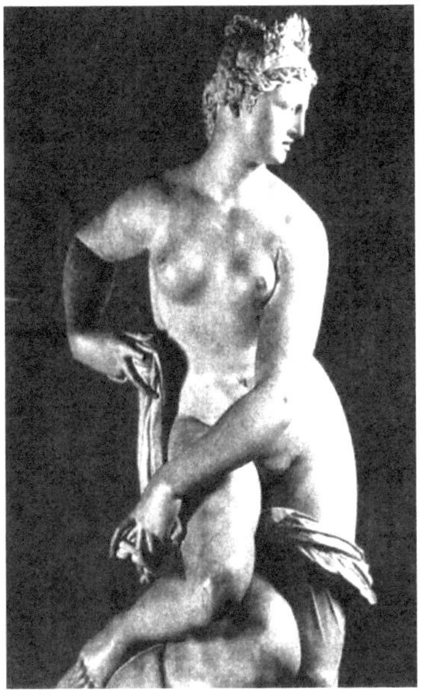

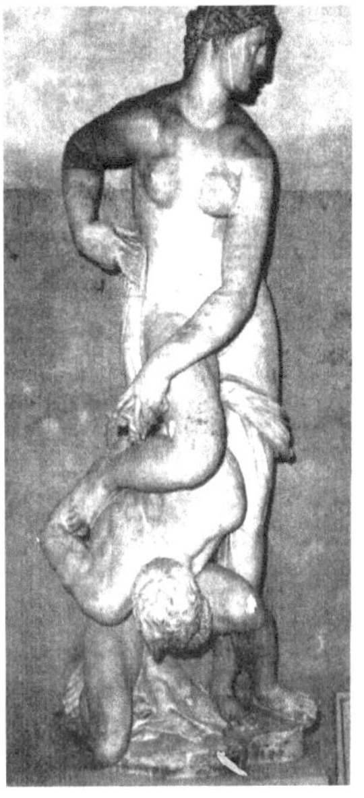

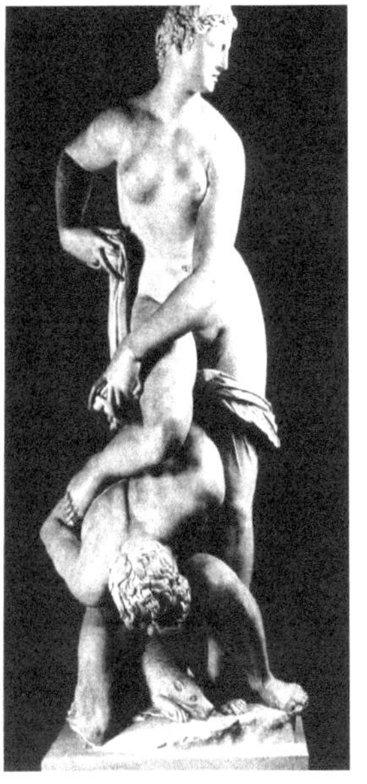

1. The flood-damaged part was washed roughly by the debris-filled floodwater but was left stained with black oil that penetrated into the most absorbent parts of the marble. After a preliminary surface cleaning with cotton wool and white spirit, a pack of sepiolite and Shellsol A was applied and repeated on the areas where the black oil had penetrated most deeply into the marble structure until all traces were removed.
2. The parts of the sculpture undamaged by the flood – the shoulders, neck and head of *Florence* – were dirty with a thin layer of black sulphation; cleaning was carried out using a pack of sepiolite mixed with deionised water.

Once the sculpture group was fully dry, the right big toe of *Pisa* was reattached by 'bonding' with Sintolit. Protection of the whole sculpture was achieved by an application of a thin film of cosmolloid wax (microcrystalline synthetic wax) mixed with white spirit; the excess was removed and left to dry. It was then polished and finally dusted with French chalk (talcum) to complete the work.

The pietra serena *relief of* Madonna and Bambino *by Jacopo Sansovino*

The relief was covered with polluted mud that had penetrated into the pores of the pietra serena, leaving small black stain dots all over the surface of the artwork. The cleaning of the mud was carried out with deionised water without touching the gilding and the heavy oil stains were removed very carefully with methylene chloride. The painted area was consolidated with MS2A diluted with Shellsol A.

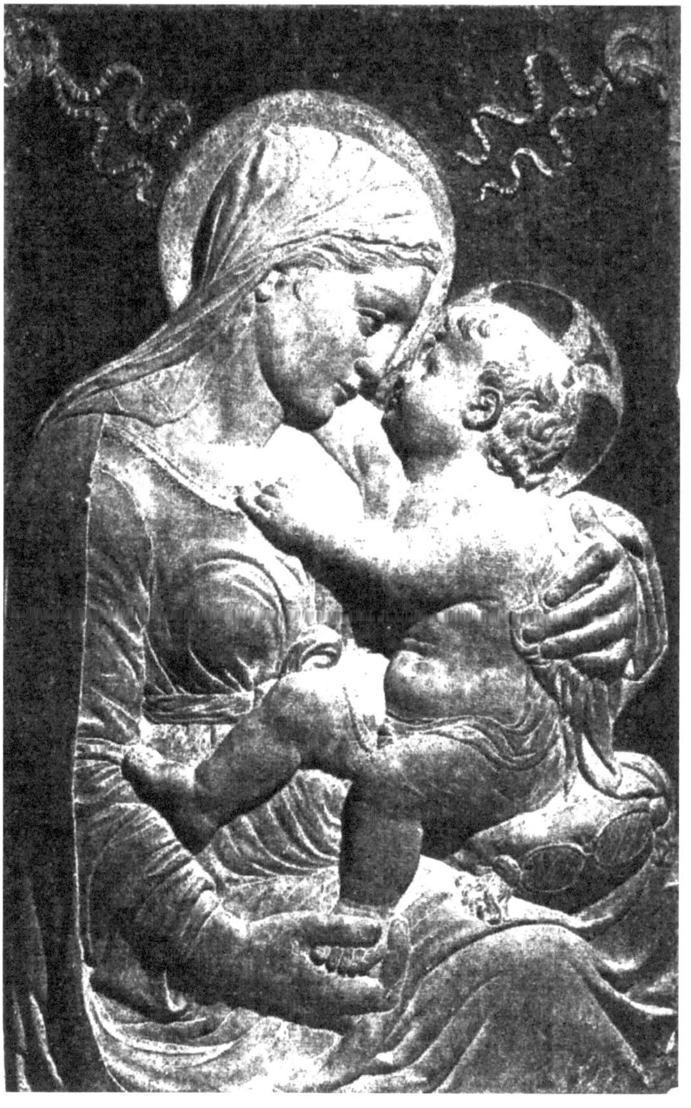

The terracotta model of Honour Triumphant over Falsehood *by Vincenzo Danti*

The model had been shattered by the force of the flood into over 100 pieces and was considered beyond repair, but after mud was cleaned from the fragments, the surfaces of the terracotta were consolidated with polyvinyl acetate diluted with water. When completely dry, the larger pieces were bonded with Sintolit and the smaller ones with PVA emulsion adhesive. This difficult task was facilitated by studying a bronze copy. It was necessary to insert stainless steel dowelling to reinforce and rejoin the right leg of *Honour* and also the calf of the right leg of *Falsehood*. Some missing fragments were replaced with Polyfilla mixed with pigment. Minor fillings were dealt with in the same way.

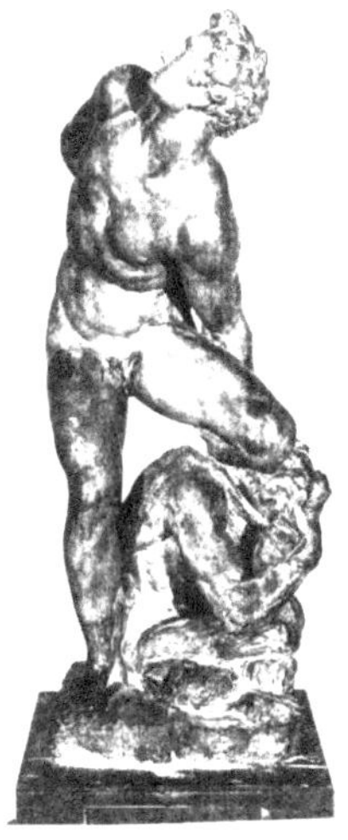
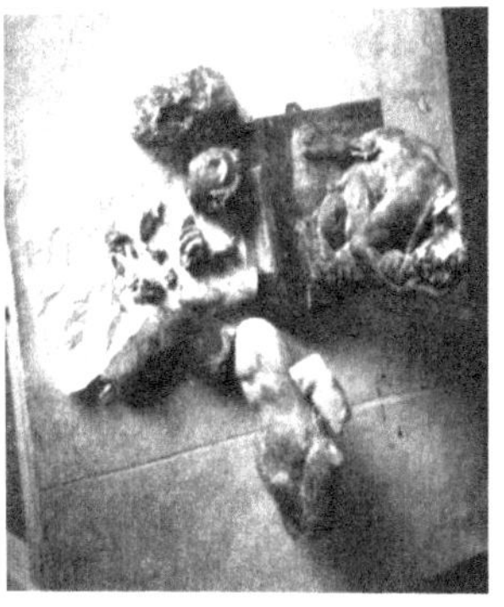
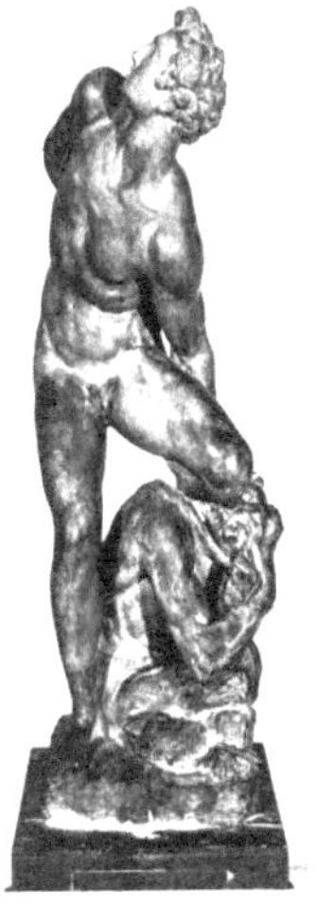

The unbaked clay with wax finish model of The River God *by Niccoló Tribolo*

The piece was broken up by the swelling of the clay armature as it lay soaking in the floodwater: 75 pieces were found among the flood sludge. The mud was removed from each piece with a soft paintbrush and deionised water. When fully dry the large pieces were bonded together with Sintolit and the smaller ones with PVA emulsion.

All the cracks between the wax coverings were filled with Polyfilla and retouched to the wax finish with dry pigment and MS2A.

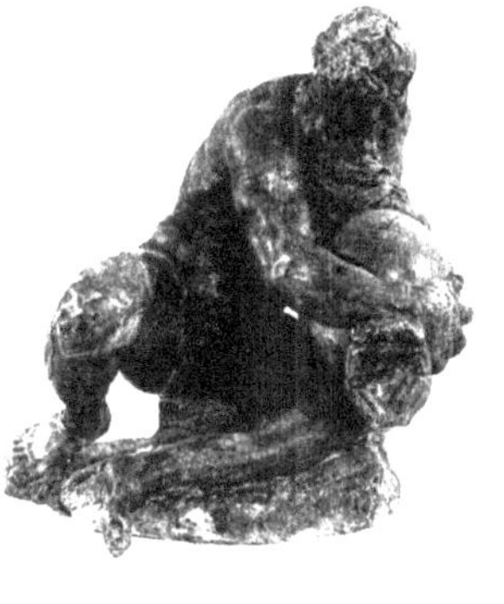
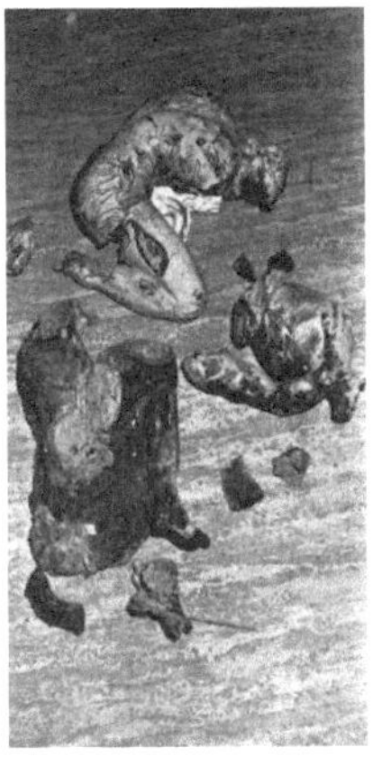
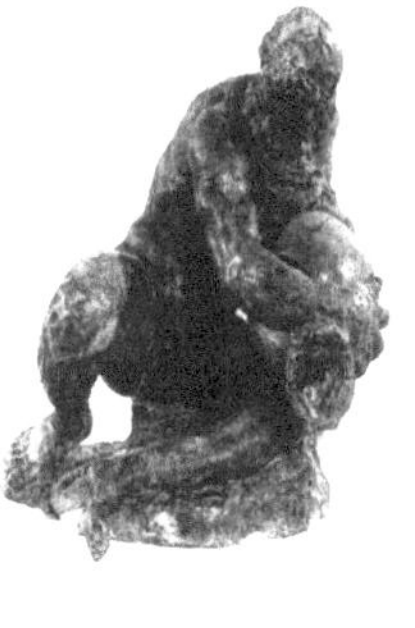

Note

1. In the words of the then superintendent of fine arts, Ugo Procacci, writing a dedication in a book he donated to me: 'To Kenneth Hempel with the most heartfelt gratitude for the precious collaboration on the rescue of the artistic patrimony of Florence, 3rd March 1967.'

New methods of paintings conservation developed in response to the flood

Andrea Rothe

The flood caused some of the greatest damage in history to the Florentine art patrimony. The city was invaded by over 450,000 tons of mud and detritus. Many paintings, sculptures, furniture and other art objects were destroyed or severely damaged. The heating oil that leaked out from the many storage tanks stained everything, including the walls inside and outside of the buildings. According to the official report, the damaged panel and canvas paintings amounted to 734 in the museums and churches alone, not counting those in private collections and art galleries. Eleven fresco cycles, 39 individual scenes and 31 frescoes that had previously been detached were covered with the mud and black heating oil; 144 sculptures in public collections of which 22 were made of wood, innumerable archeological objects, furniture, frames, and many others were also damaged.[1] We will never know the exact number of all the artworks damaged or destroyed, but it certainly could be in the thousands. The famous Cimabue crucifix in the Museum of Santa Croce was one of the most severely damaged of the many great masterpieces in Florence.

The Museum of Santa Croce, which is below the level of the Arno, suffered the worst damage. Another severely damaged area was the restoration studio of Leonetto Tintori in Via della Ninna, on the ground floor of the Uffizi, where I had worked the year before. I then moved to Palazzo Pitti. I had completed the restoration of a painting by Domenico di Michelino (a follower of Fra Angelico), *Madonna and Child with Six Saints*, and it was ready to be returned to the church of St Stephen in Anghiari when disaster struck. The studio was completely flooded up to the ceiling and the panel toppled into the mud. It swelled two to three inches in the 36 hours that it was submerged. When we gently lifted it up, virtually all the colour had become detached and the fills I had made were all that was left on the panel. Today the painting is in one of the storage facilities of the Soprintendenza for the 'incurables'.

Three statues, part of a group of the *Adoration of the Magi* from the museum in Fabriano, that were also in Tintori's studio, were jammed up against the vaulted ceiling by the rising waters. Barbara Schleicher, a private restorer in Florence who had been cleaning them before the flood, spent many months afterwards just laying down blisters in the paint film with wax. Eventually the blistering was stabilized, and she was able to remove the rest of the extensive overpaint. The restoration was completed after about three years, and the three statues were sent back to Fabriano. This is an example of a very successful treatment by just one restorer responsible for the same work of art, as opposed to many works of art treated by a team of restorers.

All the state-owned panel, canvas and wall paintings were under the jurisdiction of the superintendent, Ugo Procacci. Umberto Baldini, director of the restoration department, was put in charge of emergency procedures after the flood. All the damaged paintings treated by the restorers of the Soprintendenza were first sprayed with nistatine, an antibacterial solution with fungicidal action, against the extensive mould growth, and then faced with tissue paper and Paraloid B-72 resin dissolved in xylene or toluene. About 250 of the panel paintings were placed in the Limonaia, the large greenhouses in the Boboli Gardens behind Palazzo Pitti, and Baldini managed to secure special funding to adapt the large space and have sophisticated equipment installed to control the humidity levels.

At first the paintings were exposed to very high levels of humidity that were gradually reduced in an attempt to return the paintings to more normal conditions. The idea was grandiose, and it was thought to be the only way to deal with such a large volume of waterlogged paintings at one time. Unfortunately, as Baldini himself writes, each individual panel reacted differently to the gradual reduction of the humidity levels.[2] Many behaved erratically and deformed even more than in their initial condition when they were rescued from the water and mud. The panel paintings shrank beyond their original size causing giant blisters and tenting of the paint layers. The many layers of facing with Japanese tissue and acrylic resin created veritable cardboards on the surface of the paintings. Paraloid B-72[3] was used exclusively for this; it is a harder and less pliable resin than Lucite 2044,[4] which was also proposed. Later, however, conservators encountered considerable difficulties in their efforts to soften and dissolve away the facings.

Robert Feller, a prominent expert on varnishes and solvents, who arrived in Florence after the flood, recently recounted that he had learned that, very early on, someone had started the rumor that Lucite 2044 would crosslink and therefore become insoluble. Apparently, that was the primary reason it was set aside and not used. The truth is that unless Lucite 2044 is exposed to far more intensive daylight than is ordinarily encountered in a museum, the process of crosslinking would be very slow – possibly taking as much as 50 years before it became difficult to remove. Since the facings were only intended to serve for a few years, it is possible that the more easily soluble Lucite 2044 would have proven to be more suitable than Paraloid B-72 (also called Acryloid). Paraloid B-72 should exhibit no tendency to crosslink, but regrettably the difficulties encountered in the removal of B-72 facings have never been fully investigated nor has a technical explanation been provided.[5] Unfortunately, after 40 years there are paintings that still retain the old facings such as Giorgio Vasari's *Last Supper* from Santa Croce. Today a better choice might be the use of other synthetic varnishes such as Regalrez, which are soluble in mineral spirits and are much less toxic than the solvents required to dissolve B-72 (xylene, methyl-ethyl-ketone and toluene). In retrospect, one thing that we have learned is that not all flooded paintings should be faced indiscriminately. Instead, trying to lay down the blisters with an adhesive as soon as possible, and repeating the procedure as new blisters are formed, seems to be the better approach.

One of the biggest problems was cleaning the surfaces of the frescoes and statues that were coated with brown heating oil. Trying to clean them with solvents would only have created more problems and driven the staining deeper into the wall or stone. As a result, far more successful methods were developed using different types of poultices such as gypsum, silica or attapulgite[6] with an aromatic solvent. The applications had to be repeated many times to be effective. The same attapulgite and solvent poultices were used on stained frescoes in combination with nistatine to impede the growth of bacteria. Incidentally, the use of poultices after the flood has totally revolutionized the cleaning of frescoes. Today most cleanings are done with poultices, in some cases using just distilled water. These combinations are more effective than the conventional washing with sponges and water which was standard practice before the flood.

On the basis of new findings, I would advocate the drying of water-soaked wood as soon as possible after a flood. Leaving wood to dry out gradually seems to worsen the deformation, as tests completed in the laboratories of the Fortezza da Basso have shown.

Since 1177, when the first floods or *piene* of the Arno were documented, there have been 56, eight of which were ruinous. Many paintings such as the *Madonna delle Arpie* by Andrea del Sarto in Palazzo Pitti bear clear signs of past floods. After 1966, plans were drawn up to avoid such disasters in the future. It was recommended that works of art should not be exhibited or stored on the ground floors of the museums that suffered the greatest damage such as the Uffizi, the Museo di Storia della Scienza, the Accademia, the Museo Archeologico, the Museo dell'Opera di Santa Croce, the Biblioteca Nazionale Centrale di Firenze, the Museo del Bargello, the Museo Bardini, and many others. Flood control plans were drawn up and many were implemented. The most ambitious called for a canal to be constructed to divert part of the waters of the Arno in case of a flood, but was never carried out. Others, however, were completed. For example, the project to control the waters of the Sieve river was completed some years ago with the construction of the Bilancino reservoir, which regulates the amount of water that feeds the Arno in case of a torrential downpour. The Scolmatore of Pontedera, a large overflow basin to protect Pisa, was finished in 1987. The Arno was dredged out to lower the riverbed and accommodate a larger volume of water. Today, the embankments and the riverbed are supposed to be cleaned twice yearly.

The Ponte Vecchio, which probably has Roman foundations and is situated at the narrowest point of the Arno, had always posed the biggest problem because the openings between the arches and the stone shelf below restrict the flow of the river. Twice in the past, the bridge collapsed – it was last rebuilt in 1345. During the 1966 flood, the biggest fear was the collapse of the Ponte Vecchio, which shook violently with the enormous quantity of water pushing on it. Had the bridge collapsed, it would have created a giant dam causing even more devastation.

With the immense volume of water pushing downriver, even the modern Amerigo Vespucci bridge farther down became a partial barrier. Fortunately, in 1980 work was completed to lower both the underwater stone shelf of the Ponte Vecchio and the Ponte Santa Trinita. Both the dams of La Penna and Levane (upstream on the Arno) that had been blamed for part of the disaster are better controlled today. Rain sensors have been placed every 8 km along the Arno, and the time span to forecast a possible downpour is now less than 24 hours.

One of the positive aspects of the flood is Baldini's legacy of convincing the ministry in Rome in 1967 to create the Florentine Centre for Conservation, called Opificio delle Pietre Dure at the old Medici *fortezza* or fortress (Fortezza da Basso), with its own superintendent. This institution has since dealt with most of the 'problem children' from the *alluvione* and has become highly successful in recovering many of the most severely damaged paintings and sculptures. It has also become a conservation centre for all of Italy, and great works are treated there such as Mantegna's S. Zeno altarpiece.

Over 40 years have passed and unfortunately complacency has again set in as it had in the past. Sooner or later such a flood could hit again. In the meantime, many works of art, rare books, archives, archeological objects and historical scientific equipment are again housed on the ground floor, and the situation is ripe for another disaster. The severely damaged Cimabue is now again in the Museum of Santa Croce as it was before the flood, but now it is suspended by a pulley system so it can be pulled up to the ceiling in an emergency. An important exhibition of eight restored masterpieces damaged in the flood opened in 2006 in the same museum that was flooded 40 years ago by more than 18 feet of water, mud and heating oil.[7]

Hopefully, the commemorations of the flood (and this symposium) will raise the awareness that the great flood of 1966 could very well happen again. A plan was presented on 18 February 2005 to regulate the waters of the Arno by creating a wider riverbed as was done successfully along the Po river many years ago. But, as we have seen in Venice with the *acqua alta*, where a plan to control the high tides after 40 years still has not been implemented – these plans could take much longer than the estimated 10 years to carry out.

The flood in Florence has taught us that taking preventive measures is the best way to protect museum objects. We should expect disasters like this everywhere. The flood has also taught us to ask questions, and employ the best available materials and methods such as, when facing, to use adhesives of greater solubility. Wood should not be kept artificially wet after flooding, but dried off as soon as possible. Constant observation of a flooded object by the restorer, and treating blisters as soon as they form on a painted surface is much more effective than letting the panel paintings dry out first. Facings are not always the solution because one cannot monitor what is happening to the paint surface. The nistatine that was used was very effective in keeping the paintings from rotting. The poultices, like the attapulgite ones that were introduced to remove the heating oil residues on the frescoes and sculptures, revolutionized the cleaning of frescoes. Floods and fire are definitely the worst things that can happen to works of art, but severe acts of vandalism cannot be ruled out. We all love and admire the superb works of art that are part of the great heritage of Florence, and we must remain constantly vigilant to preserve them for future generations.[8]

Notes

1. *Rapporto sui danni al patrimonio artistico e culturale*, C.E. Giunti, Florence 1967, p. 10.
2. *Idem*, p. 64.
3. Paraloid B-72, made by Röhm and Haas is a copolymer of ethylmethacrylate and methacrylate, and requires a 'stronger' (more polar) solvent to dissolve it, such as a pure aromatic solvent like toluene.
4. Lucite, made by the DuPont Company (also called Elvacite in Italy) is available in three types: 2044, the softest, is a normal-butyl methacrylate polymer, soluble in a petroleum or 'odorless' solvent and 20% of an aromatic hydrocarbon such as toluene; the harder one, 2045, an isobutyl methacrylate, 2046 is medium hard, which essentially is a copolymer of both.
5. Robert Feller, pers. comm. October 2006 and May 2007. I am very grateful for Robert Feller's reading of my paper and his valuable comments.
6. Attapulgite, also known as palygorskite, is a magnesium aluminum silicate, a type of Fuller's earth. It is derived from a clay found in the south-eastern United States, particularly in and around Attapulgus, Georgia, after which this type of clay is named.
7. *Angeli, Santi e Demoni*, edited by Marco Ciatti, Cecilia Frosinini and Chiara Rossi Scarzanella, Edifir, Florence, 2006.
8. I would very much like to thank my wife Jeanne McKee-Rothe for her input, careful reading and editing of this paper.

The flood and the Palazzo Davanzati laboratories

Kirsten Aschengreen Piacenti

It seems impossible today, but in 1966 the Soprintendenza of Florence had no chemical laboratory and no conservation laboratory for the decorative arts. The flood highlighted this need and through the international relief funds, a temporary solution was found. The end results are the specialised laboratories in the Fortezza da Basso. They are known the world over, whereas the intermediary stage has been forgotten. It is my intention in this paper to relate what happened and who was involved: to remember a heroic moment and perhaps the finest example of international collaboration.

At the time of the flood, I was living in Pisa where my husband was associate professor of chemistry at the university. I had a research scholarship from the Consiglio Nazionale delle Ricerche (CNR) and a contract from the Ministry of Culture to catalogue the ivories in the Museo degli Argenti at Palazzo Pitti. I commuted every day to Florence from Pisa. We had just moved into a flat in a *palazzo* on the Lungarno Gambacorti, and on the morning of 4 November we watched the swelling river from our second floor windows. The situation in Pisa was very different from that of Florence, since Pisa was used to occasional floods. There were boards ready, called *palafitti*, which the military (*genio civile*) affixed to the high parapets along the river and backed with sandbags as soon as the level of the water looked threatening. As the situation got worse, the military would go into the country upstream from Pisa and open the dykes thus flooding the fields – in spite of protest from the farmers who were out with shotguns.

The river reached the top of the *palafitti*, and here and there splashed over, flooding the ground floor levels; it was a mighty torrent and the mass of water nearly filled the arches of the bridges. Large tree trunks beat against the walls – even a rowing boat, I remember. The real damage to Pisa, however, was revealed months later when the Solferino bridge suddenly collapsed and the *palazzos* on the northern side of the river had to be evacuated as their foundations were undermined by the impact of the water. But that day, while we were shocked by the violence in front of our eyes, we heard a rumour that in Florence things were much worse – that the whole city had been flooded.

It was hard to believe. My husband was a Florentine born and bred, and still had an old aunt living in the centre. We could not get through to her on the telephone, so we took the car (that just shows how little we knew of the situation). We had to leave the car in Piazza San Marco as there was no way of getting beyond. The water had subsided by then, but there was mud everywhere and the city sewers were open. Franco's aunt of course would not budge and so, much shaken, we returned home.

But the next day I managed to get back (by bus as the railway line was flooded) and put myself at the disposal of the Soprintendenza. Palazzo Pitti where I was normally 'affiliated' was too high up to be touched by the flood, the water reaching only as far as the top of Via Guicciardini. The Bargello on the other hand had suffered a great deal, so I offered my help to Professor Luciano Berti, who at that time was the director of both the Bargello and the Palazzo Davanzati museums.

The situation in the Bargello was simply unbelievable. The water had reached the level of 4.20 metres. The museum had an important library and all the books were buried in mud. With the help of volunteers (*angeli del fango* or mud angels) we tried to clean them and put them on the upper floor to dry. However it soon became clear that the greatest damage was not done by the water alone but by the oil deposits from the flooded central heating systems which settled like black smudges on the statues in the courtyard. And there was nothing we could do to remove them. We needed chemists to advise us and three were found: two from the University of Florence (Professors Ferroni and Speroni), and my husband, Franco Piacenti, from Pisa.

The Bargello also contained valuable wooden and leather artefacts, arms and furniture. Furthermore, not only the Bargello, but the whole city was full of damaged decorative art objects, and the Soprintendenza had no laboratory in which to restore them. Florence has always been famous for its artisan workshops, but they were the ones that had suffered most from the water. So we conceived the idea of setting up a temporary set of laboratories on the third floor of Palazzo Davanzati. The modification work obviously had to be undertaken by the technical office (*ufficio tecnico*) of the Soprintendenza – but where to find the money and the know-how to carry out the conservation?

Then one dark evening in Piazza degli Uffizi, I happened to run into my old professor from London, Nicolai Rubinstein, who had taught me medieval history during my first year at the Courtauld. He was the leading advisor to the British relief fund and was walking in the rain with its chairman, Sir Ashley Clarke. In the course of our conversation I told them about our pressing needs and Berti's and my ambitious project to meet them. Not only did they they offer to help but they also promised to mention it to the other relief funds for it was quite obvious that an undertaking of such magnitude could only succeed if tackled by a joint international force.

All the relief funds rallied! The American Committee to Rescue Italian Art (CRIA) provided the funds for the setting up of laboratories for the conservation of sculpture, and a small chemical laboratory to which the German fund (under Dr Kreuter) also contributed. The British fund donated complete laboratories for the conservation of arms as well as wooden artefacts. The Scandinavian countries sent three head conservators to coordinate the work of their relief funds – Leif Plahter, Steen Bjarnhof and

Erling Skaug – each in turn spending a year in Florence. They were all frequent visitors, and advisors, to the Davanzati laboratories. Indeed all the relief funds sent whatever experts we requested.

The first person they contacted was Sir John Pope-Hennessy, then director of the Victoria and Albert Museum in London. The most urgent task, after all, was to do something about the statues and their oil smudges. And it just so happened that a conservator at the V&A had already found a method to solve that kind of problem. Kenneth Hempel had invented a system of applying inert earth and solvent to 'suck' the oil from marble, the famous 'Hempel Pack' as it soon became known.

The third floor of Davanzati was adapted in record time, and as the first of its new inmates, Kenneth, arrived, I was there to receive him. It was dark and cold and there was no central heating. I opened the door in a wall niche and produced a bottle of cognac – to succour him in case the cold became too extreme. I am glad to say that he did not turn round on his heels and flee! On the contrary, he immediately set to work and was soon applying his 'Pack' to all the statues. We had found three young sculpture restorers to help him: Guglielmo Galli who unfortunately died some years later, Angela Camargo who is still working in Florence, and Lorenzo Rossi. They in their turn were helped by the volunteers, the mud angels.

Norman Brommelle, head of the scientific laboratories of the V&A, came over several times to assist with advice and suggestions. In the meantime a small chemical laboratory was set up in one of the rooms, under the direction of Norman's wife, Joyce Plesters, who was then head of the chemical laboratory of the National Gallery. She stayed for a month and was succeeded by Unn Simonsen, the wife of Leif Plahter, who occupied a similar position at the National Museum of Oslo. These two highly professional ladies shared their specialised knowledge with two Italian chemists, Valerio Malaguzzi-Valeri and Raffaella Rossi Manaresi, both good chemists but who had never thought of using chemistry in the service of art.

The laboratory set up by the American relief fund, CRIA, was headed by Bill Young from the Boston Museum of Fine Arts. He advised on the conservation of such problematic items as the painted leather shields and reliquaries of varying materials. The actual work was carried out by two paintings conservators, Kay Silberfeld from Baltimore and Maddalena Saiko from Vienna.

Bill asked for an infrared spectroscope which CRIA provided. Professor Gilmore (or was it Millard Meiss?) was troubled by the high cost and asked me to ensure that the instrument would not be relegated to a deposit at the end of the emergency. He need not have worried – it was taken over by the Centre for the Study of Causes of Decay and Methods of Conservation which, in 1974, my husband was able to set up within the framework of the Consiglio Nazionale delle Ricerche (CNR) (National Council of Research) and in collaboration with the University of Florence.

Right from the beginning, Dr Otello Caprara had been called in from Bologna to help direct the conservation of the enormous quantity of damaged wood artefacts. These were essentially the sacristies and stalls of the flooded churches and had to be restored *in situ*. But there were many other smaller objects that were treated in the Davanzati laboratory – even the *Mary Magdalene* by Donatello, from the Baptistery was there for a period being climatised after the long spell in the water before being placed in the care of the paintings restorers at the Fortezza. Caprara was aided by two Florentine artisans, Giuliano Arretini and Orazio Fossati, who remained with the Opificio after the transfer of the laboratory.

Danish restorer Arnold Hansen helped with the veneered furniture, and two Swedish restorers each spent a year restoring the great relief by Grinling Gibbons (Fig. 1). This was normally housed in the Bargello, but at the time of the flood it was at Bearzi's bronze conservation laboratory opposite the Uffizi, where it was half submerged in the water. As it happened I was there to fish it out and collect the pieces that were floating about. Professor Procacci and Caprara did not want to have anything to do with it – 'it is nineteenth century,' they said – and left it there. They would not listen when I told them that it had been commissioned from the artist by Charles II and in 1682 donated to Grand Duke Cosimo III.

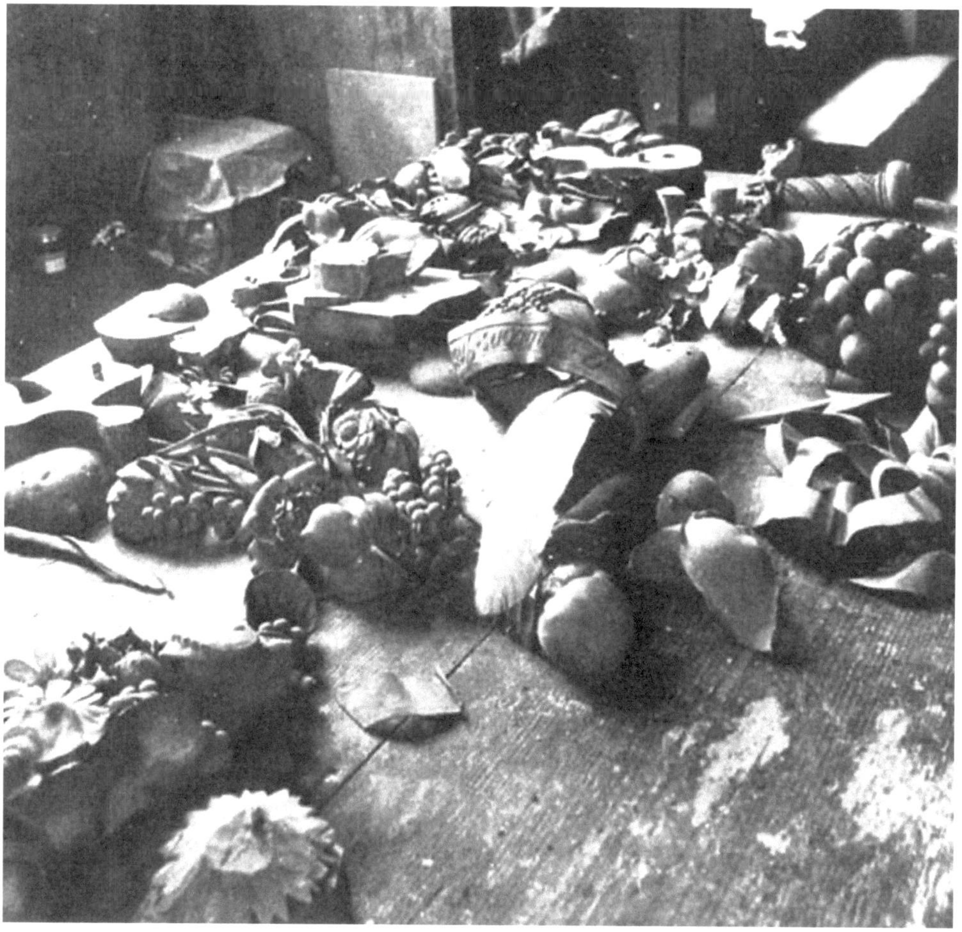

Figure 1 Grinling Gibbons, trophy before conservation.

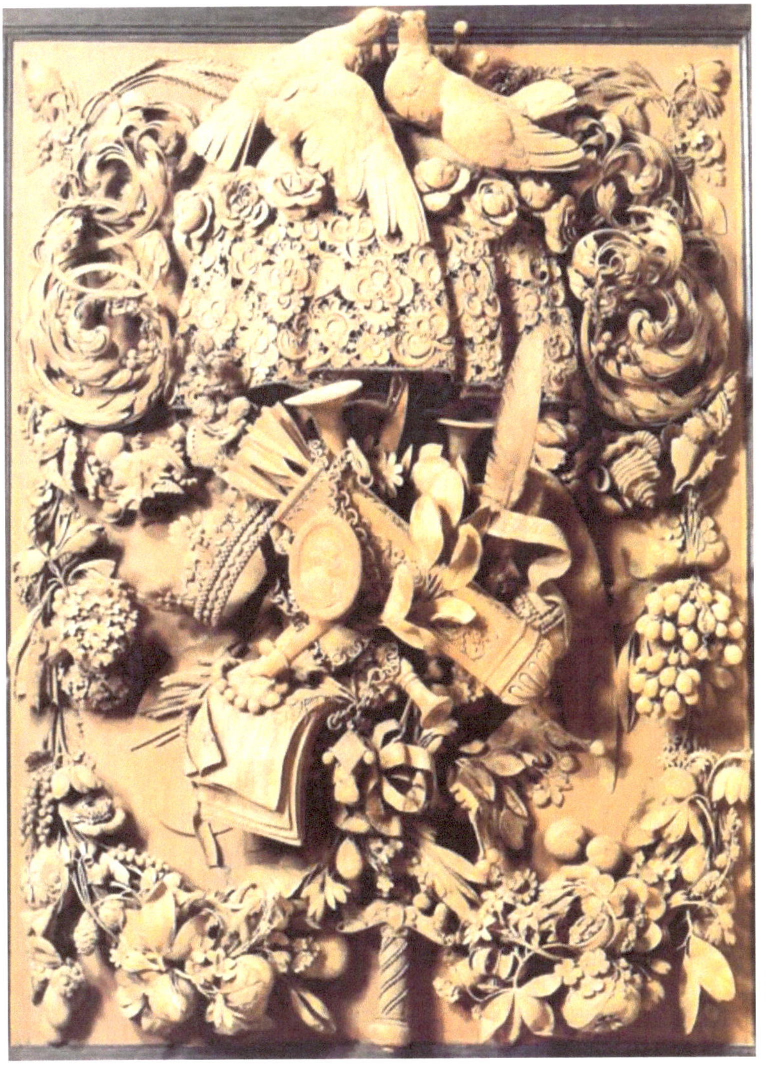

Figure 2 Grinling Gibbons, trophy after conservation.

Following its restoration by the Swedish restorers, it was shown in the exhibition *Firenze Restaura* of 1972 (Fig. 2). It was then sent back to the Bargello where it was placed under a leaky window, so it was soaked again. Then I asked for it to be transferred to Palazzo Pitti where it was hung over a door in the *grotticina* which, for a short while, was my office, and there it was at the time of the gas explosion – which devastated both me and the room – and it got blackened. Finally it was loaned to the V&A (and restored) where, in spite of all these unfortunate vicissitudes, it was the centre of the great Grinling Gibbons exhibition of 1998 and hailed as his *chef-d'oeuvre*!

The arms of the Bargello were put under the particular care of Lionello Boccia, a connoisseur of arms and armour and later director of the Stibbert Museum. Most of the pieces from the armoury had been cleaned of mud immediately after the flood (under the supervision of Emma Micheletti, vice director of the Uffizi) and were then sent to Vienna to be restored at the Waffensammlung under the guidance of Bruno Thomas, the director. But 826 objects still remained and of these, some were restored at Davanzati by Heinz Dauer from Germany, helped by Florentine Paolo Nencetti, who remained at the Opificio as metal conservator.

At the Bardini museum the water level reached 3 metres and submerged the showcases containing 25 musical instruments, which were all seriously damaged. The American conservator Bernard Rabin and his collaborators gave them first aid and later, two German experts, Olga Adelman from Berlin and Friedeman Hellwig from Nurnberg, took care of the actual conservation during several visits to Florence. Moreover they chose the five pieces to be sent to Holland for restoration. This generous offer was the result of a benefit concert (*serata di beneficenza*) specially held in a school for this purpose. One of the helping mud angels was Alan Wilcox who remained in Florence and who is now a highly esteemed lute maker.

At the end of 1967, a small exhibition was held at the Bargello of the works of art which had by then been restored: *Mostra di restauri a sculture e oggetti d'arte minore.* The catalogue is the one remaining witness to the existence of the laboratories in Palazzo Davanzati. By 1972, at the time of the exhibition *Firenze Restaura*, the memory had dimmed almost to extinction. That exhibition was held to celebrate the pre-flood conservation laboratory and only a few sentences here and there in the catalogue mention the contributions from abroad.

The Davanzati laboratories were dismantled in 1968 and incorporated in the Soprintendenza conservation centre, together with the Italian restorers and chemists. This meant that from then on, the Florentine Soprintendenza had laboratories for the conservation of sculpture, wood and metal, with fully fledged restorers to operate them. But the greatest change was in the field of science versus art – experience of the flood had taught a new respect for the use of 'science for conservation'.

In 1974 the CNR with the University of Florence set up a research centre under the direction of my husband (he became full professor of organic industrial chemistry in 1968). Stone decay was studied, and new methods and materials for the protection of stone were tested. Collaboration took place with research laboratories abroad, for instance with North Carolina State University and DuPont Industries. Later, university courses at all levels were set up in Florence and today have a wide following. It is indeed a completely different world where finally scientists and art historians speak together.

It was the end of my involvement with conservation and, I thought, with CRIA. But there I was wrong. Some years later I myself entered the Florentine Soprintendenza as the director of the Museo degli Argenti. One day Marco Chiarini, my colleague from the floor above (he was the director of the Galleria Palatina), asked me if I could think of a use for the newly restored Casino del Cavaliere at the top of the Boboli Gardens. This is a small garden pavilion built as a retreat for the grand dukes and embellished in the eighteenth century. It stands in its own rose garden and is flooded with sunlight from huge French windows. It had been offered to Marino Marini, the sculptor, in exchange for donating his works of art to the city of Florence, and was restored with the idea of it becoming a private residence with central heating, kitchen and bathroom. Then, when it came to the point, the citizens of Florence refused to give it Marini. So the embarrassing problem arose of what to do with the restored building.

I suggested turning it into a porcelain museum, porcelain being about the only material that suffers neither from light nor from heat. The remains of the grand ducal porcelain collection had so far been preserved in a mezzanine of the Museo degli Argenti and would benefit from an airier location. The only trouble was that the collection had never been systematically studied and I knew nothing about porcelain. And then, strangely enough, CRIA came to my aid!

CRIA had instituted a series of scholarships to allow young American students to work for a while in the Florentine museums that had received so much money from the USA. The Museo degli Argenti had not suffered from the flood as I mentioned earlier, but all the same, one day an American voice rang out under the loggia outside my window: "Hey there, here I am!!!" That was Sheila Tabakoff, a student of porcelain who had embarrassed the CRIA organisers – they did not know where to send her until they heard of my pressing need for just such a person.

She was a wonderful collaborator – knowledgeable, cheerful, hard-working – who arrived wearing the shortest possible mini-skirt, which positively shook our staid world where mini-skirts were as yet unknown. With her help, we managed to organise the collection, dividing it up into nationalities and manufactories. We found the missing inventories from the Savoy period so that the history of the collection could be researched. Svend Eriksen (whom I knew from Waddesdon Manor in the UK) arrived to compile the first catalogue of the Sévres porcelain.

On 8 September 1973, the Porcelain Museum was opened. It was a glorious day, sunny and warm. The mayor came one hour too early as his secretary had muddled up the times. We were still cleaning the showcases, so my husband took him round the rose garden. And then the guests arrived – the whole city. It was magic. It could not have been done without the help of the CRIA scholar Sheila – and that period in Florence was the beginning of her career in the world of American porcelain museums.

I should like to end by expressing my gratitude to CRIA and the other relief funds for their wonderful support in times of need.

References

Mostra di restauri a sculture e oggetti d'arte minore, catalogue of the exhibition
Museo Nazionale del Bargello (December 1967–February 1968), Firenze, 1967.

Armi storiche del Museo Nationale di Firenze, Palazzo del Bargello, restaurate dall'Aiuto Austriaco per Firenze, catalogo di Bruno Thomas e Lionello G. Boccia, Firenze, 1971.

Firenze restaura: il laboratorio nel suo quarantennio, guida alla mostra a cura di Umberto Baldini e Paolo Dal Poggetto. Mostra di opere restaurate dalla Soprintendenza alle Gallerie (Fortezza da Basso), Firenze, G.C. Sansoni, 1972.

Transfer of panel paintings after the flood

Erling Skaug

The Centro Nordico del Restauro a Firenze (Nordic Centre) was in operation for three years (1967–70) with participants from Finland, Sweden, Denmark and Norway. One hundred conservators of paintings, furniture and books contributed altogether over 34 years of work.[1] The length and continuity of our presence allowed us to undertake long-term tasks such as transfers of panel paintings. We were also able to follow parallel works, for example that on the large crucifix by Cimabue at the Fortezza da Basso.[2]

Transfer is a very drastic intervention. Encountering the expert tradition of Italian *trasporto* was for us a fascinating first-time experience. As I have not been involved in transfers since that time, lack of supplementary experience limits any further thoughts I might have on the subject. Nor have I been able to examine the transfers done at the time to ascertain their present state in concise scientific terms. The qualified judges in retrospect must be our Florentine colleagues themselves.[3]

The transfers we carried out were supervised by Gaetano Lo Vullo, Edo Masini and Vittorio Granchi. One of them was a *Madonna and Child with Two Saints* by Neri di Bicci (1474–75).[4] The procedure is described in some detail since this kind of intervention is hardly to be found in textbooks and can rarely, if at all, be taught in training programmes today. A few variants of the method will also be mentioned and comments given where appropriate.

Preliminary treatment

The painting, measuring 75 × 127 cm (Fig. 1), had been completely submerged. A provisional front protection of tissue paper and Paraloid B-72 was applied soon after by the Florentines, and the painting was kept under high relative humidity in the specially installed storerooms in the Limonaia of the Boboli Gardens. In March 1968, it was

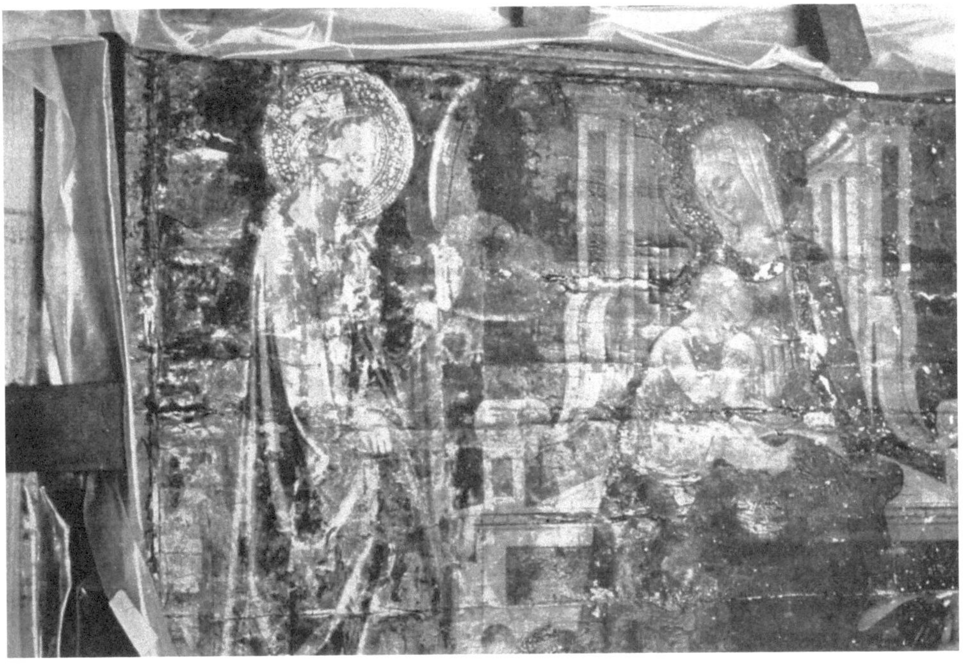

Figure 1 Neri di Bicci, ***Madonna and Child with Two Saints and Donor***, resting on wet sawdust and wrapped in PVC foil during treatment at the Fortezza da Basso, Florence.

transferred to the Fortezza and entrusted to the Nordic Centre. The treatment lasted for two years and was finished in May 1970.

During a previous treatment, long before the flood, the mechanical strength of the panel had been reduced. Grooves had been sawn from the back, parallel with the wood grain, to about half of the thickness of the panel. This was one among several methods traditionally employed to correct warp in the wood. Due to its fragile state, the painting was handled on a stretcher all the time.

Before transfer, the provisional facing had to be removed, paint and ground with their many blisters fixed, and a new facing applied. The inevitable loss of moisture made the panel shrink and new blisters were forming continuously. This behaviour, in addition to the poor state of the support in general, left us in no doubt that a transfer had to be undertaken.

The panel had to be kept in a swollen state during the preparatory treatment. In the beginning, it rested on a bed of wet sawdust, wrapped in PVC foil; this caused severe attack by mould in spite of repeated sprayings with thymol (Fig. 1). As an alternative, the back of the painting was sprayed with water every morning, and cups with hot water placed underneath and changed at intervals. At the end of the day it was again sprayed and laid down on the cool brick floor, which had been moistened in advance. Cups of hot water were placed around the painting, and everything was covered up for the night by a tent of PVC.

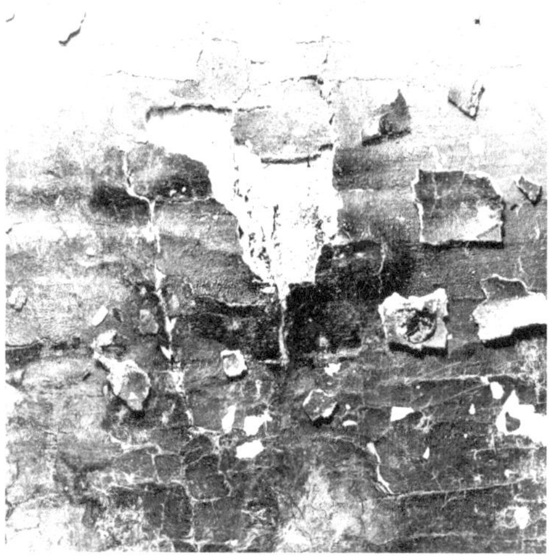

Figure 2 Detail of Figure 1. Collapsed blister, to be reconstructed and, later in the process, laid down and flattened.

A basic problem was that the shock-like expansion of the wooden support during the flood had led to plastic deformation (stretch) in the layers of paint and ground. When the panel subsequently began to dry and shrink, the paint and ground could not follow. The pictorial area had become larger than its present support, making it impossible to put down the largest blisters. This factor alone made transfer the most realistic solution.

The largest blisters had collapsed (Fig. 2) and were reconstructed as 'domes,' filled with gesso from a syringe, and their flattening left to a later stage of treatment. The fixation work progressed parallel with the removal of the provisional facing. Cotton swabs were soaked in xylene or methyl-ethyl-ketone and applied on the surface in areas of about 10 × 10 cm at a time, and covered with aluminium foil. After 10–12 minutes the paper usually loosened from the surface.[5] A difficulty encountered in the process was that the Paraloid appeared to become less soluble with time, requiring longer solvent contact. This was surprising, given its reputation of high stability. Another problem was the failing effect of the rabbitskin glue used for the consolidation of paint and ground. This was tentatively attributed to the paint layers' possible content of greasy fuel oil mixed with the floodwater, but repeated extractions with xylene, toluene and methyl-ethyl-ketone resolved this problem.

The potential adverse effects of prolonged exposure to solvents were considered, mainly with regard to the paint layers. Egg tempera is a resistant medium,[6] but the possible invisible effects of leaching could not be checked.[7] However, there was little choice under the circumstances: the preliminary front protection had to be removed in order to proceed with the transfer, and the increasing insolubility of the Paraloid made the operation (actually comprising a large number of paintings in the same situation) a race against time. The health effects were also considered but the ventilation trunks eventually installed did not work until later, so, after breathing in solvent vapours for months, we may perhaps be excused for possible amnesiac tendencies.[8]

Transfer

After the preparatory stage a new facing was applied consisting first of two layers of tissue paper with Paraloid followed by an application of a thick gesso layer reinforced with an open-weave cotton canvas. This 'plaster cast' would protect the topography of the surface during the subsequent stages of treatment. Next the painting was turned face down, clamped to the table on a thin foam rubber mattress, and the wood shaved down by an electric plane with minimal vibration. The remainder of the wood was removed by gouging.[9]

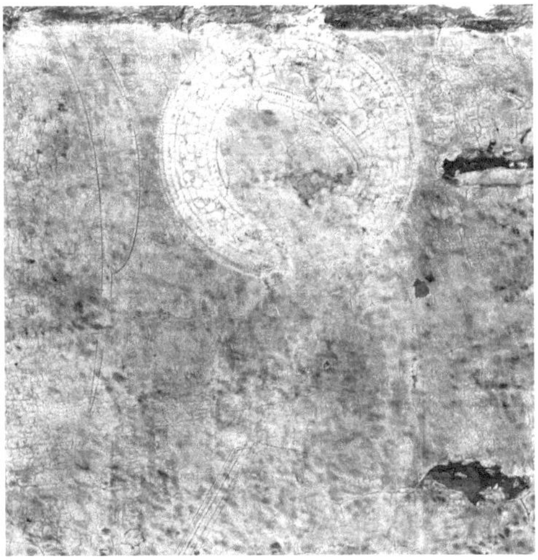

Figure 3 Detail of Figure 1 seen from the back during transfer, with incised outline of St Catherine. A thicker layer of ground has been retained in the area of the halo and crown in order to preserve the topography of the punchwork.

The ground was then thinned down. It was powdery in parts, but it adhered well to the paint layer as we approached it from the back. It was decided not to remove the ground completely, contrary to procedures carried out in many other cases, one reason being to retain the relief of the punched decoration in haloes and dress borders (Fig. 3). Since the adhesion to the paint was good anyway, there was no need to remove it and the new ground would presumably stick well to its older relative.

Transfers, like most interventions in conservation and restoration, offer unique opportunities for observations of the artist's technique. A preliminary drawing was made on the ground with a silver point or a pen. Incisions of parts of the composition, i.e. those that could be made with the compass and ruler, and the outlines of areas to be gilded, could be observed and traced (Fig. 4). Paint structures were examined in cross-sections, pigments analysed and a few infrared photographs taken; identification of Cennino Cennini's *oro di metà* was given a new point of departure.[10]

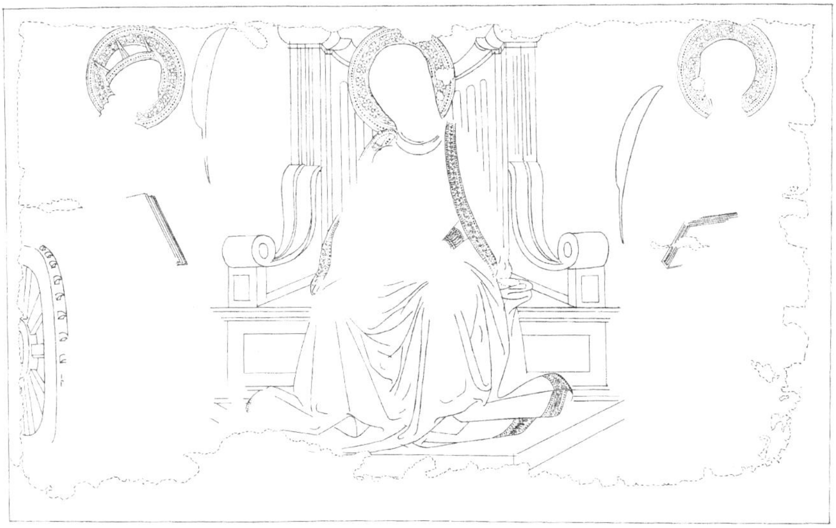

Figure 4 Tracing of the incised lines of the composition, as seen from the front.

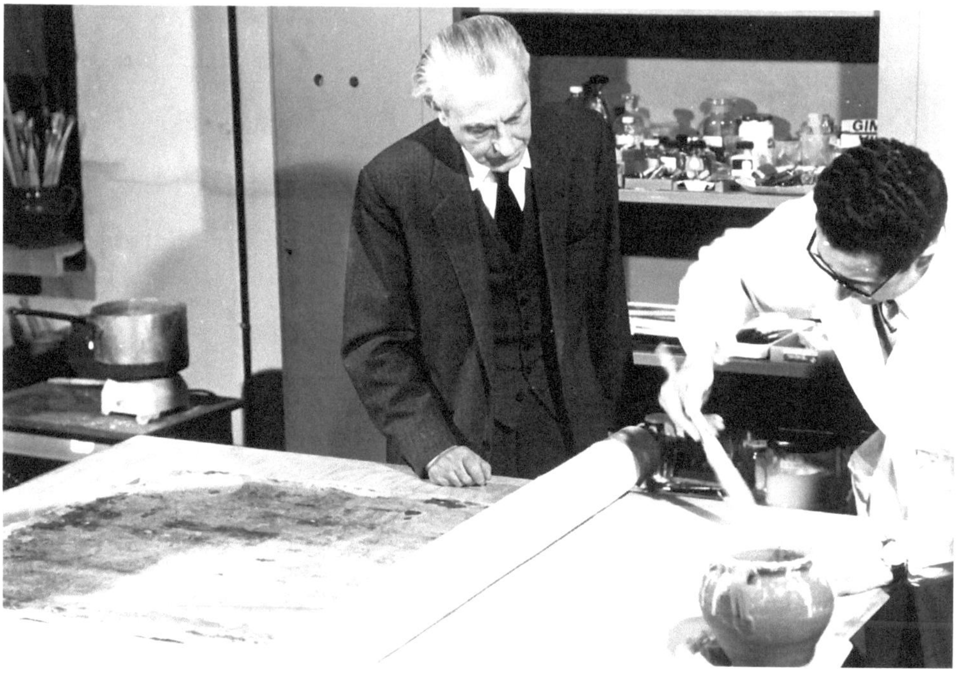

Figure 5 Gaetano Lo Vullo (right) and Edo Masini during the application of the new support. The painting is lying face down with a front protection, new gesso is rubbed thoroughly into the thin skin of the original gesso, canvas is put on from the roll, and another layer of gesso applied over it. The process proceeded in a piecemeal fashion, about 20 cm at a time, to ensure that the first layer of gesso did not dry before the second layer was applied and the canvas was well embedded.

At this stage blisters and irregularities were flattened, sometimes after further thinning of the ground. Local moistening was followed by local and total cold pressure, varying from overnight to several days and certain improvements were achieved. The final adjustments to the surface were left to a later stage.[11]

Before the new support was applied, the back of the painting was brushed with oxgall and a thin sizing of rabbitskin glue. Gesso for the new support was then applied. These stages were carefully instructed by Edo Masini and watched over by the grand old man of Florentine painting conservation, Gaetano Lo Vullo (Fig. 5). The gesso consisted of well-slaked gypsum ('gilder's whiting') in rabbitskin glue mixed with a little oxgall and molasses for flexibility, plus some pigment (ochre and burnt umber) for a visual modification of the white new ground.[12]

The new support was a thin open-weave canvas embedded in gesso, i.e. with the same structure as the front protection. Arguments in favour of a rigid support were raised: a panel painting should remain a panel painting. On the other hand future interventions would be easier to carry out on a canvas support. The thin canvas, washed and stretched in advance, was thought to have a marginal tendency to induce new, foreign crackle patterns in the painting, different from panel craquelure. With this potential problem in mind, we might today have considered a synthetic, moisture-stable canvas of, for example, polyester.

After one week of drying, the painting was turned face up and the front protection removed (Fig. 6). The painting was then cleaned, the overpaint and old fillings removed, and the lacunae filled with new putty of chalk and gelatine level to the surface of the paint layer. Further adjustments were made on the vacuum hot-table, a new tool introduced in Florence after the flood. A mixture of beeswax-dammar (7:3) dissolved in white spirit was brushed onto the painted side, the back was moistened and the painting treated several times under low pressure, alternating between face up and face down, with a rubber mat placed between the painting and the hot-table.

Finally, the back was evened out with a filler, sandpapered and brushed with oxgall. A traditional hand lining with glue and warm irons was then done by the Florentines before the painting was tacked onto a laminated panel. This last phase, carried out by my successor in cooperation with the Florentines, seems today almost like a review of lining methods north and south of the Alps up until the early 1970s: both wax and glue, both hot-table and irons. In hindsight – especially after the Greenwich conference on lining in 1974 (Villers 2003) – perhaps less would have been more.

After inpainting and varnishing, the painting was deposited in the Pinacoteca di Siena, where it can now be seen; 36 years after treatment it looks fine to the naked eye.

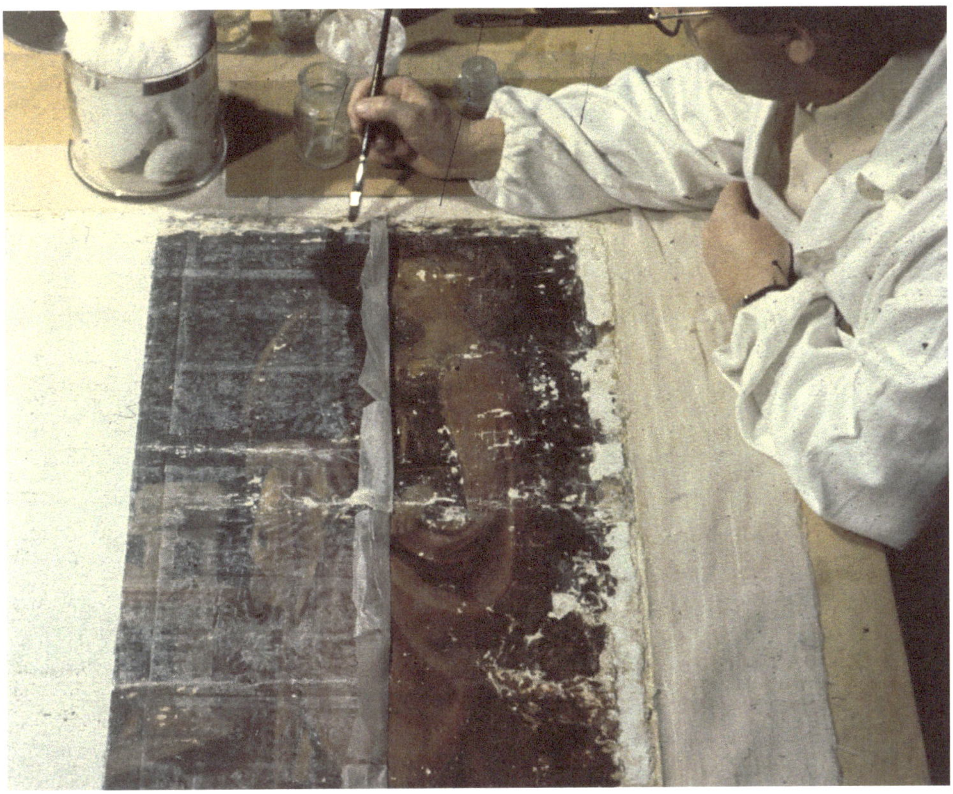

Figure 6 Removal of front protection after transfer to a new support. Left to right: canvas embedded in gesso; two layers of tissue paper and Paraloid; painting with tissue paper removed; the new support.

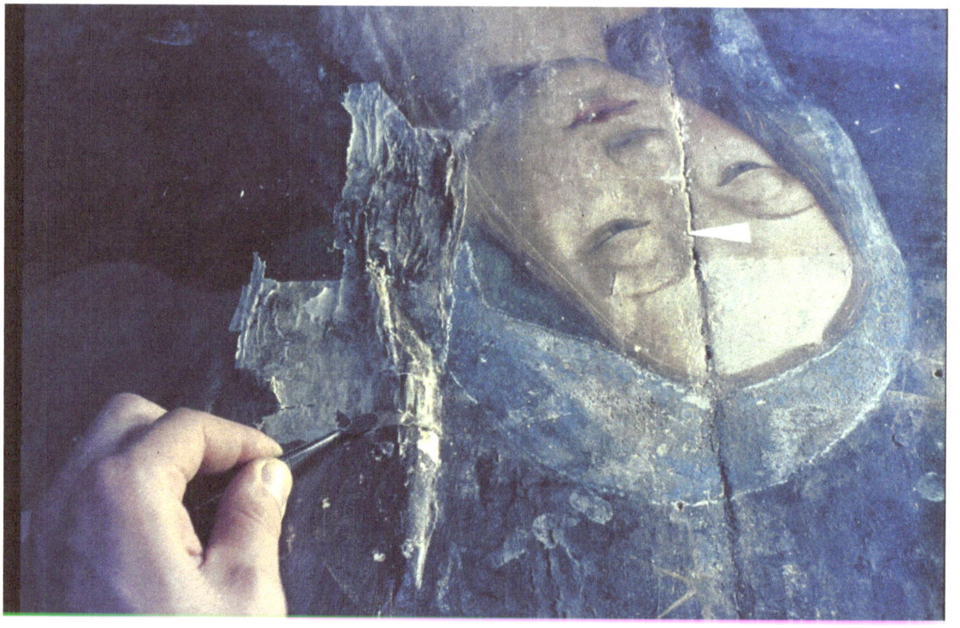

Figure 7 The *Madonna del Giglio*, from the church of San Giuseppe, during preparatory work before transfer.

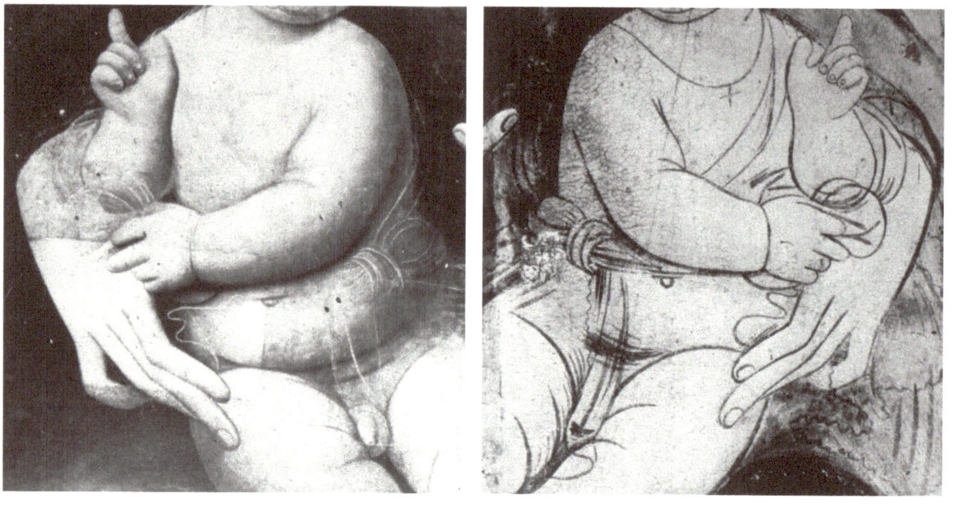

Figure 8 Detail of Figure 7 during transfer: the Child from the front (left) and from the back (right), showing the preparatory drawing for the composition.

Other cases

The so-called *Madonna del Giglio* from the church of San Giuseppe, attributed to the Master of Marradi, was a similar case. The wood was in a better condition than in the Neri di Bicci, but its successive expansion and contraction upon drying presented the same problems with large blisters (Fig. 7). The parallel processes of removing the provisional facing and fixing blisters and loose paint took a long time.

The Nordic Centre carried out the work up to the complete removal of the ground only, exposing the underdrawing made by brush (Fig. 8). Also the first layers of underpaint were revealed, such as green earth (apparently without *verdaccio* modelling) under the flesh colour. After we left in 1970, the remainder of the work was done by the Florentines.

From the point of view of the history of technology as well as of style, it is of course interesting to study the painting from the 'inside' to see how the artist began. Like the observation of incisions in the Neri di Bicci, this was an extra bonus as the work was done primarily for a different purpose, allowing rich opportunities to compare with Cennino Cennini's instructions of about 1400.[13]

Finally, I will give a brief recapitulation of the transfer of Cimabue's Santa Croce crucifix. Supports are valuable historical testimonies and we wanted to save not only the painting but also the wooden cross and its interesting construction. That meant separating the painting from the wood by a variant of the *strappo* process. A separation was possible since the whole panel had been covered with canvas before grounding – a normal procedure in the Italian *duecento* and *trecento*.[14]

By careful moistening of the glue between panel and canvas, in small areas at a time, the canvas could, little by little, be separated with a flat spatula. Blotting paper was placed in between the canvas and the support as the work proceeded. Vittorio Granchi was responsible for the work, to all ends and purposes an impressive and elegant solution (Fig. 9). The same method was used also for an altarpiece by Lorenzo di Niccolò.[15]

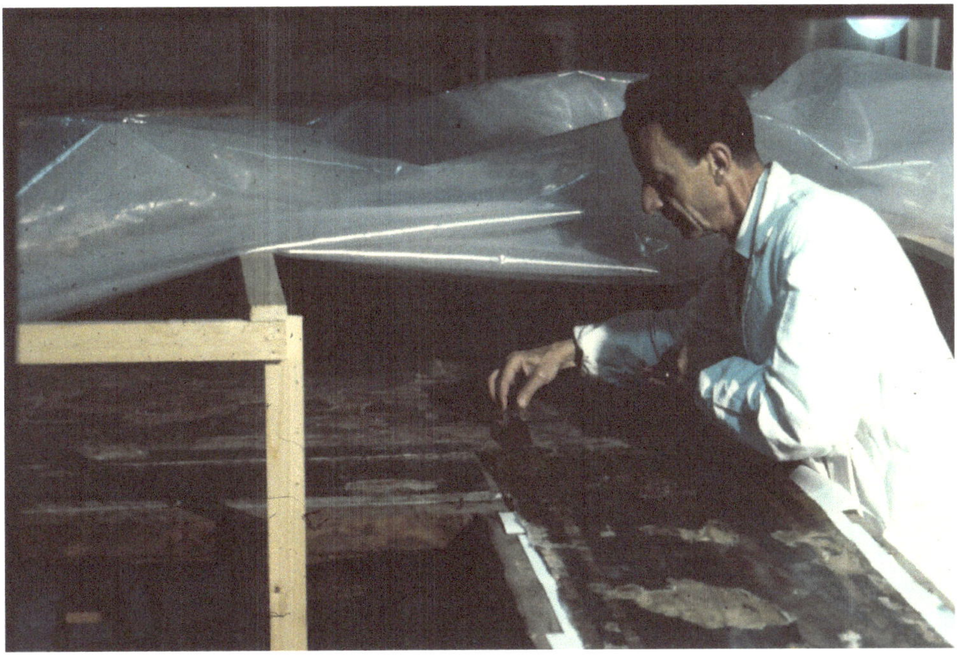

Figure 9 Cimabue, crucifix from Santa Croce. Vittorio Granchi in the course of separating the triple layer of canvas, ground and paint from the wooden support.

Another bonus for technology historians was offered when the wooden support was laid bare. For example, the joins between the main panel and the cross-arms had been filled with fibres. The purpose was apparently to make a flexible filling that would help to maintain a level surface underneath the canvas, so that the joins would remain invisible in spite of the movements of the wood. A rectangular insert, covering a knot or an irregularity in the wood, could be seen in the main panel, supplemented by a filler perhaps of sawdust, gesso and glue.

Conclusion

In many cases it was possible to consolidate the support and thus avoid a transfer.[16] To a temporary collaborator from abroad, however, the transfers that were carried out seem to have been well justified under the circumstances. On the whole, the Florentines did a great job in the wake of the flood, from the organisation of the work – including *la seconda alluvione*, i.e. the 'flood' of foreigners who arrived to help – to the critical interventions of conservation and restoration. Visitors to the Fortezza da Basso today will appreciate the changes and improvements that began over 40 years ago. With the development of new methods and the application of modern technology, the formulation of a teaching programme in conservation and an impressive frequency of publications, Florence is a striking example of how a disaster can be turned to advantage.

Notes

1. Most of the participants were employed in museums and were given paid leave from their institutions and their travel expenses covered. Each participant spent at least three months in Florence, some up to one year. A steering committee with representatives from Nordic museums, archives and political authorities was established in Copenhagen, and directors of the centre, all painting conservators, were appointed for one year at a time: Leif Einar Plahter (Oslo) for the first year, Erling Skaug (Oslo) for the second and Steen Bjarnhof (Copenhagen) for the third. Björn Hallström (Stockholm) acted as a stand-in when the directors changed. A publication planned at the end of the project unfortunately did not materialise, but some of the manuscripts were published separately; Plahter 1969, Plahter and Plahter 1971 and 1975, and Skaug 1978.
2. A first exhibition of restored works was held less than six years after the flood, see Baldini and Dal Poggetto 1972. For glimpses of the Nordic Centre and our work see figures 81 (preliminary examination only), 100, 103–4 and 119–20.
3. For the most up-to-date summary of the flood and description of a number of post-flood restorations, see Ciatti *et al.* 2006.
4. The painting was flooded in the conservation studios at Vecchia Posta, Uffizi. Bruno Santi (1976, p. 423) has identified the painting as *Tavola da Radda di Mariotto Ghondi* (considered as lost at the time of printing) with a commission date of 27 September 1474. The painting is presently on display in the Pinacoteca di Siena.
5. See illustrations of this phase in Plahter 1969 (reprinted in Skaug 1999).
6. Before treatment, the binding medium had been analysed by Unn Plahter, chemist at the Oslo University Museums, who joined the Nordic Centre during its first year. Samples were washed in xylol to rid them of the Paraloid, then tested in morpholine and in 10% KOH, which gave no reaction, indicating egg

tempera. The subsequent treatment above, including cleaning and removal of overpaint, showed a very high resistance to organic and inorganic solvents. Pure egg in this period was indirectly confirmed by two analytical campaigns published a few years later: a survey showed that some *quattrocento* paintings had alternating layers of egg and oil, others were painted in pure egg (see Johnson and Packard 1971). In a broad supplementary project a painting by Neri di Bicci was found to have been painted in egg throughout, probably with alternating layers of glair and yolk or mixtures of both (see Gay 1972).

7. With regard to leaching and 'paint starvation' in general, I refer to an informal conversation in 2003 with Alan Phenix, then at the Courtauld Institute of Art, London. Phenix's studies of cleaning methods so far have, among other things, led to the conclusion that the phenomenon of leaching – demonstrable in laboratory tests and model studies – is of less importance in practical work.
8. Also disinfestation with a spray gun, whenever it was thought possible, was often done without protection. When browsing through photographs from the project today, the lack of masks and fume extraction systems – by no means unusual for Florence at the time – strikes one as a significant difference between conservation routines then and now.
9. Strips of canvas, covering the joins between the boards of the panel, were discovered at this stage.
10. The term *oro di metà* has, in editions of Cennini's *Il libro dell'arte*, either been thought to be an alloy or just translated literally as 'half gold'. Unn Plahter's analysis of the metal leaf in Neri di Bicci's painting subsequently led to the conclusion that the term must correspond to the well-known laminated gold-silver leaf *Zwischgold* etc.; see Skaug 1978, p. 225 and note 1, Post Scriptum, pp. 234–5, followed by Skaug 1993, pp. 16–18.
11. The potential problem with laying down large blisters is the risk of deformation of the composition. The swelling and shrinking of the panel produce local blisters and the paint surface may not be returned exactly to its original shape by flattening because the surface is stretched. Whether the problem is hypothetical or merely marginal, the practical problem is in itself often considerable. In Alessandro Allori's large panels in Santa Croce, the fibres pasted locally over the joins in the panel caused enormous local blisters. The process of laying them down was carefully documented; see Baldini and Dal Poggetto 1972, figures 77–80, and Ciatti *et al.* 2006, pp. 29–31. Comparisons made before and after treatment showed no appreciable signs of deformation. The operation, performed by Edo Masini, was among the impressive practical achievements of the post-flood works.
12. In retrospect, the purpose of the latter does not seem clear. No pigment in the original ground had been observed, and there was no tinted *imprimatura* between ground and paint: Neri di Bicci obviously did paint on a bright white ground. Perhaps the idea of toning down the new ground was influenced by a policy against total cleaning, leaving some 'patina' of surface dirt etc. in which case (unfortunately not discussed at the time) it would seem somewhat irrational to introduce 'patina' also from the back.
13. For a discussion of the date of *Il libro dell'arte*, see Skaug 1993. Further evidence for its date during Cennino's documented stay in Padua (1398–1401) is furnished, among other things, by the fact that gold leaves in the treatise are beaten from the Paduan-Venetian *ducat*, not the *florin*; see Frezzato 2003, p. 16 and note 18.
14. The variants of this practice, observed during the post-flood works, caught my attention with my diploma work in mind: the examination and treatment of a Norwegian thirteenth-century altar frontal, where only strips of canvas over joints and critical points were to be found. The latter variant seemed to appear in Italy only in the fifteenth century (cf. note 9 above), followed by loose, unwoven fibres in the sixteenth century (cf. note 12 above). Supplemented by later observations, an overview of practices north and south of the Alps *c.*1150–1550 was recently published; see Skaug 2006. The subject is expanded on in a paper by the author, read at the interim meeting of ICOM-CC, Paintings Working Group, at the British Museum, London, May–June 2007 (publication forthcoming).
15. See report and illustrations in Ciatti *et al.* 2006, pp. 122–3.
16. See, for example, Ciatti *et al.* 2006, pp. 112–15.

References

Baldini, U. and Dal Poggetto, P. (1972) *Firenze restaura. Il laboratorio nel suo quarantennio.* Firenze: G.C. Sansoni.

Ciatti, M., Frosinini, C. and Scarzanella, C.R. (2006) *Angeli, santi e demoni: otto capolavori restaurati. Santa Croce quaranta anni dopo (1966–2006)*. Firenze: Edifir.

Frezzato, F. (2003) *Cennino Cennini. Il libro dell'arte*. Vicenza: Neri Pozza.

Gay, M.C. (1972) *Application de la méthode des coloration sur coupes minces à l'étude des liants de quelques peintures italiennes du XIVe et XVe siècles*, Lisbon Congress 1972 on Conservation of Paintings and the Graphic Arts (preprints). London: IIC, 705–13.

Johnson, M. and Packard, E. (1971) 'Methods used for the identification of binding media in Italian paintings of the 14th and 16th centuries', *Studies in Conservation* 16: 145–64.

Plahter, L.E. (1969) 'Nordisk center for restaurering i Firenze', *Meddelelser om Konservering* 1(7): 3–14 [reprinted with English summary in Skaug 1999, pp. 68–72].

Plahter, L.E. and Plahter, U. (1971) 'Notes on the deterioration of Donatello's marble figure of St Mark on the church of Orsanmichele in Florence', *Studies in Conservation* 16: 114–18 [reprinted in Skaug 1999, pp. 73–5].

Plahter, L.E. and Plahter, U. (1975) 'The St Sebastian tabernacle by Leonardo del Tasso in the church of Sant'Ambrogio in Florence', *Acta ad archaeologiam et artivm historiam pertinentia* (Norwegian Institute in Rome) VI: 151–62.

Santi, B. (1976) *Neri di Bicci. Le Ricordanze (10 marzo 1453– 24 aprile 1475)*. Pisa: Marlin.

Skaug, E. (1978) 'Neri di Bicci's *Madonna and Child enthroned with Two Saints and Donor*, Uffizi Gallery, Florence: material, technique and restoration', *Acta ad archaeologiam et artivm historiam pertinentia* (Norwegian Institute in Rome) VIII: 223–6.

Skaug, E. (1993) 'Cenniniana: notes on Cennino Cennini and his treatise', *Arte Cristiana* LXXXI(754): 15–22.

Skaug, E. (ed.) (1999) *Conservare necesse est. For Leif Einar Plahter on his 70th Birthday*, Oslo: IIC Nordic Group.

Skaug, E. (2006) 'The third element': preliminary notes on parchment, canvas and fibres as structural components related to the grounds of medieval and Renaissance panel paintings', in J. Nadolny (ed.), *Medieval Painting in Northern Europe: Techniques, Analyses, Art History. Studies in Commemoration of the 70th Birthday of Unn Plahter*. London: Archetype Publications, 182–201.

Villers, C. (2003) *Lining Paintings: Papers from the Greenwich Conference on Comparative Lining Techniques*. London: Archetype Publications.

The Italian Art and Archives Rescue Fund

Frances Clarke

It was Franco Zeffirelli's dramatic documentary television film that had such a compelling impact on the British public, recording as it did the catastrophic scenes of Florence drowning in the floodwaters of the Arno. Zeffirelli was of course an international figure, whose productions at the Royal Opera House in Covent Garden were tremendously admired; when he rang the chairman of the governors of the Opera House, Lord Drogheda, to ask for aid from Britain, Garrett Drogheda in turn rang my late husband, Ashley, who was chairman of the British-Italian Society at the time, and passed the charge to him. At the same time, the Italian Minister for Education (Pubblica Istruzione), Gui, made an appeal to UNESCO on 2 December which the director-general, Rene Maheu, addressed to all member nations. The British Minister for Culture, Jenny Lee, asked Ashley to chair a committee to raise funds and allocate them as required. In this he was fortunate enough to have the indefatigable support of Lady Thorneycroft (born Malagola and wife of Lord Thorneycroft) and Mrs Humphrey Brooke, wife of the secretary to the Royal Academy, both of whom worked incessantly with enthusiastic volunteers to raise funds. The name of our committee, the Italian Art and Archives Rescue Fund (IAARF), was influenced by the plea from Professor Nicolai Rubinstein, the distinguished historian of Florentine (particularly Medicean) history, who begged for the word 'Archives' to be included in the title.

The committee met at the National Gallery within a fortnight of the flood by kind invitation of the director, Sir Philip Hendy. Besides the chairman, members included Sir Frank Francis (director of the British Museum), John Pope-Hennessy (Victoria and Albert Museum), Sir Anthony Blunt (Courtauld Institute and Keeper of the Royal Collection), Sir Maurice Bowra (Oxford University), Professor Rubinstein (Warburg Institute and Westfield College, London University), Noel Blakiston (director of the Public Record Office) and Lady Thorneycroft.

By 29 January 1967, some two months after coming into existence, the fund had collected £115,000. Already the Italian authorities had tackled the all-invading flood damage with heroic courage, but the scale of the task was immense and required urgent action: nearly one million volumes were submerged under mud, oil and sewage at the Biblioteca Nazionale Centrale di Firenze (BNCF); the Archivio di Stato alone had four miles of archives submerged; some 600 paintings from the major collections had to be slowly dried and faced with mulberry paper; and frescoes and sculptures, impregnated with oil from broken heating systems, were to be treated either *in situ* or at the Palazzo Davanzati.

By the end of January 1967, the fund had sent £50,000 to the International Committee, presided over by the mayor of Florence, for immediate disbursement to the most needy institutions, and a grant was also made to the British Institute, whose library was inundated. The fund immediately dispatched tons of materials, such as blotting paper, fungicides, disinfectants, mulberry paper and dehumidifiers, as well as financing the dispatch and subsistence of experts and students in all departments of restoration work: the rescue operation for the contents of the BNCF being almost entirely British with 12 experts and 45 students (*angeli del fango* or mud angels) working at Forte Belvedere and at a power station near the railway. In the administration of this aid, the fund had the invaluable help of the British Consulate and consul Christopher Pirie-Gordon.

The British Council in London financed visits to the Public Record Office (PRO) by the head of the new restoration laboratory at the Florentine Archivio di Stato, as well as paying for visits to the National Gallery, PRO and the Victoria and Albert Museum (V&A) by chemists and restorers, while the Institute of Archaeology helped in restoration of a sarcophagus. Joyce Plesters from the National Gallery spent two months at the Palazzo Davanzati Centro di Restauro training two chemists, and Kenneth Hempel, chief sculpture conservator at the V&A was released by the director to work at the Bargello on oil-impregnated sculpture (Giambologna). Technicians from the PRO went to help in the Archivio di Stato and picture restorers were sent at the request of Professor Procacci.

By the end of July 1967 it was decided that fund-raising activity should be wound down gradually. An account of expenditure up to 4 November 1967 details amounts paid to the Accademia delle Arti e del Disegno, the Archaeological Museum, Archivio di Stato, BNCF, Palazzo Davanzati Centro di Restauro, Horne Museum, Soprintendenza delle Gallerie and Museo della Storia della Scienza – besides which, there were the substantial payments through the British Consul for subsistence to visiting experts/student volunteers. The Archives Sub-Committee was set up as a separate ongoing operation, with the British Museum experts preparing a research programme.

Meanwhile UNESCO called an international conference to meet initially in Florence on 9–12 November 1967 then in Venice on 13–14 November. By that time, IAARF was already sending funds to Venice (£25,000) to start restoration in the fourteenth-century Gothic church of the Madonna dell'Orto, the beginning of another chapter, another fund born out of the IAARF – the Venice in Peril Fund – which, together with some 26 committees from other nations, is still operating under the auspices of UNESCO more than 40 years on.

Building a network of support for conservation: the Committee to Rescue Italian Art

Fred Licht

The mechanisms that initiated the Committee to Rescue Italian Art (CRIA) and kept the American contribution to post-flood work going for nearly a decade were as surprisingly simple as they were effective. The original impulse came from the Department of Art History at Brown University (Providence, Rhode Island) and spread rapidly throughout the academic, archival, and librarian network of America. Within two days of the flood in Florence, Professor Fred Hartt and I were in Florence to survey the damage, meet with Professor Procacci to photograph as much as we could of the worst hit monuments, and to bring back all the documentation available. On our return to New York, a press conference was held at which every major newspaper, radio, and television network was present. At the same time, the visual documentation was taken in hand by a team of cinema technicians who spliced the documentation we had brought back into an impressive film that was shown in all the first-run movie houses in American cities. All of this work was done by students, librarians, archivists, and instructors on a voluntary basis. The pump had been primed by a donation from Edward Warburg to cover the cost of our trip to Florence, the rental of an auditorium capable of handling a TV press conference and setting up a rudimentary office in New York. From that moment on, the funds began to arrive in ever larger amounts. Students, schoolchildren, ordinary citizens who had been to Florence or Venice and had fallen in love with these cities, ordinary citizens who hoped one day to visit Italy – they all contributed and contributed generously.

One or two glamorous events may have helped publicize our efforts. I remember with particular pleasure a benefit exhibition organized by Wildenstein under the patronage of Jacqueline Kennedy consisting of rarely seen Italian paintings from museums as far afield as Nebraska and Florida. Yet the actual financial results of such manifestations were proportionately small compared to contributions by private individuals. Far more important were the lectures held by art historians, historians, librarians, and archivists who toured all the most important cultural centers from the east to the west coast. I don't remember the exact amount that was contributed to CRIA after a lecture I gave at the university there, but I do recall being staggered by the generosity of the undergraduate and graduate students.

From the very beginning, all the founders of CRIA agreed that we should take on non-glamorous projects (although several fell to our lot such as the consolidation of the entire floor of San Marco in Venice) leaving the more appetizing restoration projects to those agencies that needed that kind of publicity. We were also able on occasion to take over projects begun by other agencies. For instance, when the French restorations of Santa Maria Maddalena dei Pazzi could not be completed due to lack of funds, CRIA completed the job.

The actual amount accumulated by CRIA, though large, would certainly not have allowed us to continue work for nearly a decade. Unlike other national contributions to the post-flood rescue work, we were not permitted to disburse any money directly to the agencies of another state on pain of losing our tax-exempt status. We were limited by law to pay only for completed pieces of work that we had inspected and approved. Thus the lump sum could be invested to generate interest year after year.

I would like to take this opportunity to pay tribute to one of the most extraordinary and least celebrated heroes of the post-flood era: the late Otello Caprara. Not only was he responsible for rescuing Donatello's *St Mary Magdalene*, but with unrivaled energy and vision he also trained young men and women to help with the painstaking work of rescuing splinter after splinter of precious *intarsias* at Santa Croce and other sites, while at the same time teaching them to become fully fledged experts in wood restoration. Of all the many men and women it was my privilege to work with during those years, it was certainly Caprara who was the most inspiring.

Experiences from a selection of disasters from 1966 to 2003

Alan M. Farancz

When the city of Florence was ravaged by the waters of the Arno in November 1966, the great masterpieces of Western civilization came under the most severe of assaults. Who was the first to respond? It was museum staff at all levels. It was the students. It was the people. The government came later. As we all know, water is deleterious to the survival of most art and artifacts: whereas fire and man-made disasters are finite, water is insidious in its effects on paintings, sculpture and paper objects. The time and patience necessary to bring these back to their former glory is enormous.

I entered the Conservation Center of the Institute of Fine Arts, New York University (NYU) in September 1966. The flood in Florence occurred barely two months into the program. At that time there were not many trained conservators available from either apprenticeships or graduate programs in the United States. Art students and other students as well as restoration organizations and facilities (both public and private) in Italy came to the rescue. Students from around Europe came to perform the necessary rescue and triage. A group of conservators in the US, together with art historians, traveled to Florence in December of 1966. A plan was formulated to create CRIA (Committee to Rescue Italian Art). Bates Lowry and others, with the support of Jacqueline Kennedy, joined the conservation effort by sending materials and individuals to Florence. It was decided that the students who wanted to go from the Conservation Center, as well as art history students from NYU, Columbia, and Harvard, would get travel grants to Florence in the summer of 1967.

I arrived in Florence at the end of May 1967 and was given a small apartment near Piazza della Signoria. At the office of the Superintendent of Monuments I met the other students and Bernard Rabin, who was to be in charge. I was placed in the Palazzina Meridiana of the Palazzo Pitti and given the task of removing the facing paper from two large candelabra with angels. I was shown what to do and how to reattach the flaking paint. I worked from 7 am to 6.30 pm – as I was not used to the siesta I continued to work through lunch. This was very different from a classroom situation as I worked

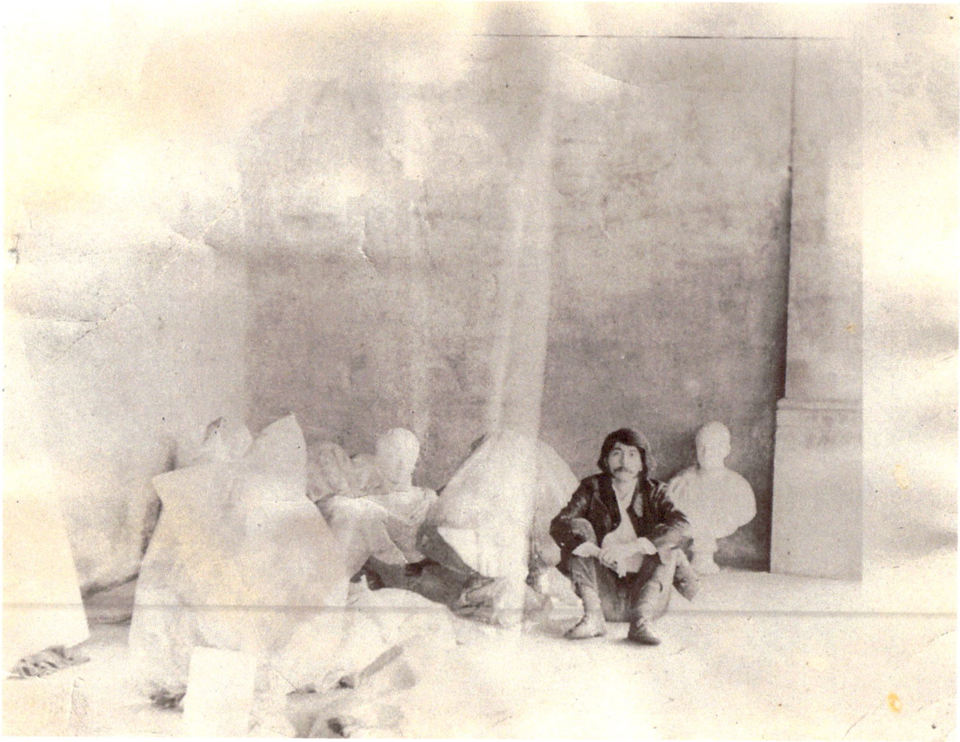

Figure 1 At the Palazzo Davanzati (1967).

totally in the apprentice mode, which made the craft element of conservation very clear to me. As instructed, I examined the sculptures to see how they were fabricated and took small samples of the wood, gesso and paint to examine later. Mr Rabin would come by once a week to discuss my progress and answer any questions that I had. After some six or seven weeks, my assigned work was finished.

Next, at the Palazzo Davanzati, where I was introduced to Dr Otello Caprara, I was instructed on how to poultice and wash marble busts (Fig. 1). I worked on cleaning some eight to ten marble busts covered with dirt and oil residue, and I was told that if I wanted to stay through the next year, a grant would be provided – but this was not to be.

I returned to New York and began my second year of classes. In the early spring, a member of the Italian Cultural Mission arrived at the Conservation Center with the paperwork that indicated I had been awarded a Fulbright Research Fellowship to go to Florence. I returned to the city in the spring of 1968 where I was assigned a project under the guidance of Dr Caprara at the Fortezza da Basso. I examined a fascinating fifteenth-century crucifix: the hair was made of rope that had been soaked in plaster and the loincloth was real fabric soaked in plaster and molded and shaped by hand to make folds. The upper body was made of wood covered with a cloth that was stiffened with a glue and chalk mixture and then painted. I met individuals who had participated in the rescue, recovery and triage of the artworks in 1966 and I saw for myself the effects walking through the Fortezza and some of the repositories for the flood-damaged artworks. It was an experience I will never forget.

I became friendly with the Centro Nordico (Nordic Centre) people who were working on a St Sebastian wood figure, as well as some conservators working on panel paintings and canvas paintings. Everyone was more than willing to talk about what they were doing. I learned the process of glue lining and of *pasta* lining; I learned a technique for cleaning off linseed oil from panel paintings; and I learned how to make gesso and how to make fills in losses. I talked to the panel and painting restorers about the pros and cons of using the new synthetic varnishes that were being developed in the US and the UK. They pointed out that the old methods had been proven over time and asked, 'what if you coated a painting with a new synthetic varnish and it changed in a way that it could not be removed?' I thought that this was a very interesting question.

When I returned to the US I was contacted by Bernard Rabin. He sent me clients and then I started working for him part-time in 1974. In time we became partners and, when he retired to join Princeton University as its conservator in 1979, I assumed the leadership of the studio.

In most parts of the world, people live surrounded by their society's artwork: the buildings, their contents and the monuments. America is a relatively young country and most people regard artwork as being that which they see in pictures in magazines, on TV, or in their local historical museum – it is the contents of the local historical museums and historical houses that are usually at risk. From 1970 to the present, I have seen dozens of disasters in museums, in warehouses, in homes and with clients. From 1990 to 2003, I have been involved in rescue, recovery and triage of every major disaster in the US, dealing with fires both small and large, interior floods caused by plumbing problems, climate control failures, hurricanes, earthquakes and major floods.

Following the Loma Prieta earthquake in San Francisco in 1989, a small Mexican museum applied for assistance for their artworks to the Federal Emergency Management Agency (FEMA) and was denied. However, some politicians worked their magic and a memorandum was filed. I was contacted by a consultant to FEMA and asked to go to Washington and help in the development of protocols for the repair of damages to artwork – an important step in ensuring that the historical fabric of our society is assisted after a natural disaster.

In Washington there seemed to be confusion about the role of the federal government in a declared disaster with regard to relief for damaged historic or artistic material. There was much discussion of whether or not they should act like an insurance company. By showing them both easily recognized works of art in the National Gallery of Art, like Picassos and Rembrandts, and then taking them to the Smithsonian to see Martha Washington's shoes and dresses, I was able to convince them to recognize these materials as important culturally and being, essentially, the fabric of the United States. Accordingly I was asked to write the protocol to cover these materials. This historic decision changed the outlook of FEMA in regard to art and artifacts damaged in a declared disaster.

My work with FEMA was essentially to identify that artworks were, in fact, artworks and then to determine if their damage was a result of the disaster or a pre-existing problem. At FEMA, I was appointed as a subcontractor under an overall contract for disasters. I presented disaster reports on all of the objects deemed to be art and identified conservation specialists to give estimates for repair (I did not take on any of the conservation work for my studio).

Much has been done towards preparedness in the past 40 years, from elaborate disaster plans, disaster wheels and disaster kits to new drying technologies for paper items and courses in disaster mitigation. I still firmly believe that however good it is for an institution to be prepared, it is still the individuals who will roll up their sleeves and put on their boots that make the difference. It is very important that those who lead the efforts have gotten wet and muddied before.

Earthquake damage, Los Angeles, 1970

From the time I spent working at the Palazzo Davanzati laboratory in 1967, I had learned how to clean marble by the poulticing method. The head of the conservation department at the Los Angeles County Museum of Art asked me to work on a damaged marble frieze (Fig. 2). As there was no tub or any washing facilities, I took it outside on the balcony to poultice it with talc and then wash the poultice off. Although it began to rain, this did not interfere with the cleaning – there are times when you just have to get the job done!

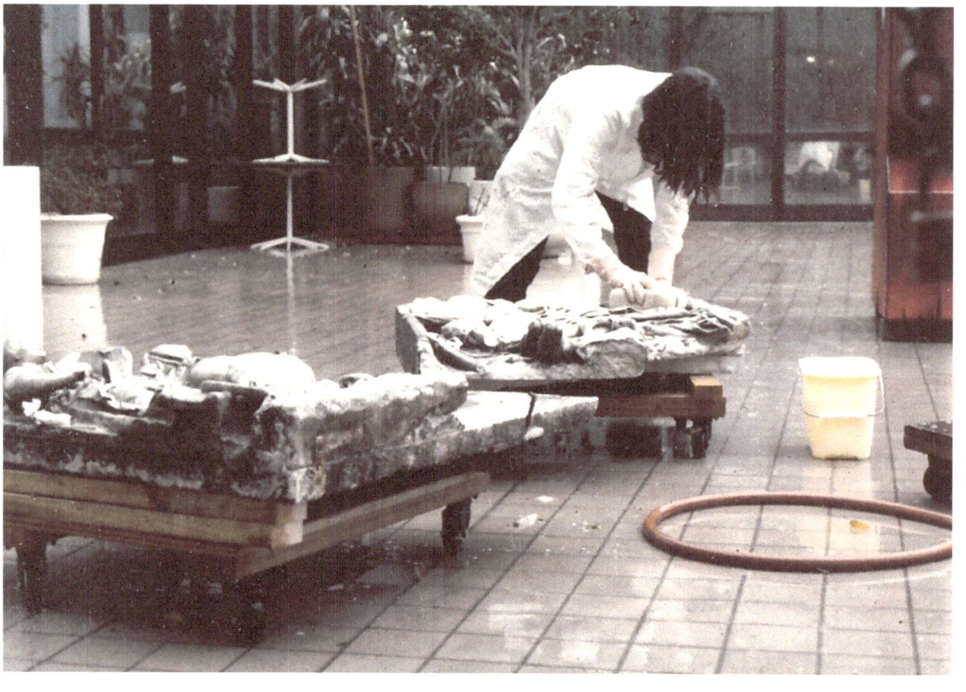

Figure 2 Working on a damaged marble frieze at the Los Angeles County Museum of Art (1970).

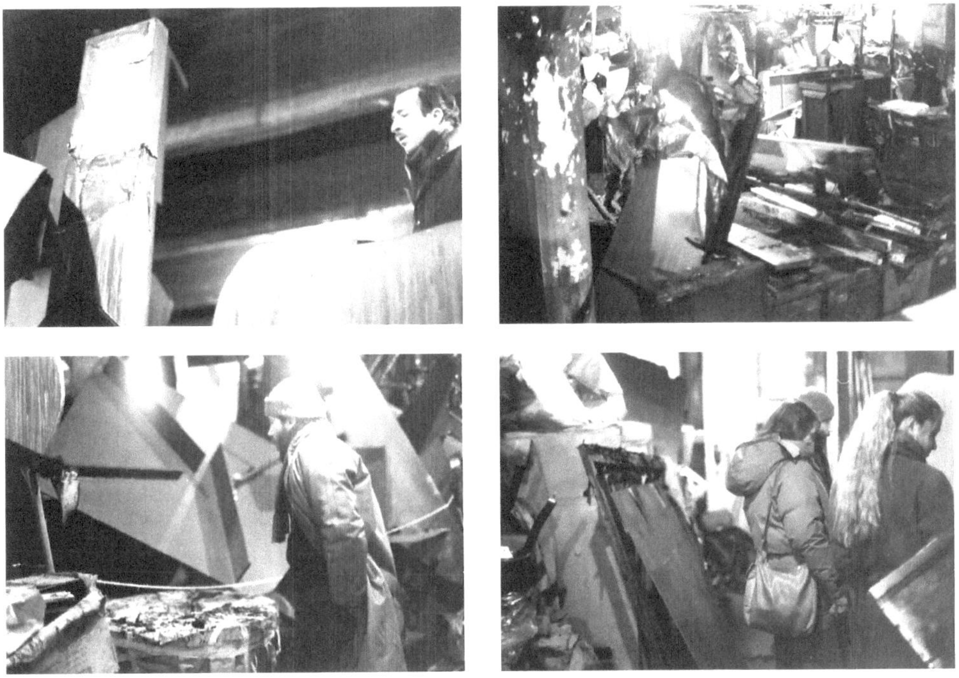

Figure 3 Art warehouse fire, New York City, January 22, 1982.

Art warehouse fire, New York City, January 22, 1982

This fire happened at about 11 pm in the middle of the winter. There were several large, painted steel sculptures by the American artist David Smith, and a few hundred stained canvas paintings by Color Field artists such as Morris Louis, Jules Olitsky, and Kenneth Noland, also clients of ours. We went in to examine the site the morning after and made some interesting observations – all the sculptures were wrapped in polyethylene and the polyethylene had carbonized onto the paint layers. We worked on this fire-damaged material for 10 years. Two Pop Art paintings in this fire were severely damaged and illustrate the range of conservation problems that might be encountered as a result of this kind of disaster: one was burned to a crisp, the other severely damaged by water (Fig. 3).

Hurricane Andrew, Florida, 1992

One disaster caused by Hurricane Andrew (Fig. 4) was at the Pioneer Museum in Florida City, a small house museum containing local history artifacts and documents. It contained everything from the small town district ranging from a football uniform from the state championship to photographs and old newspapers. The house received the

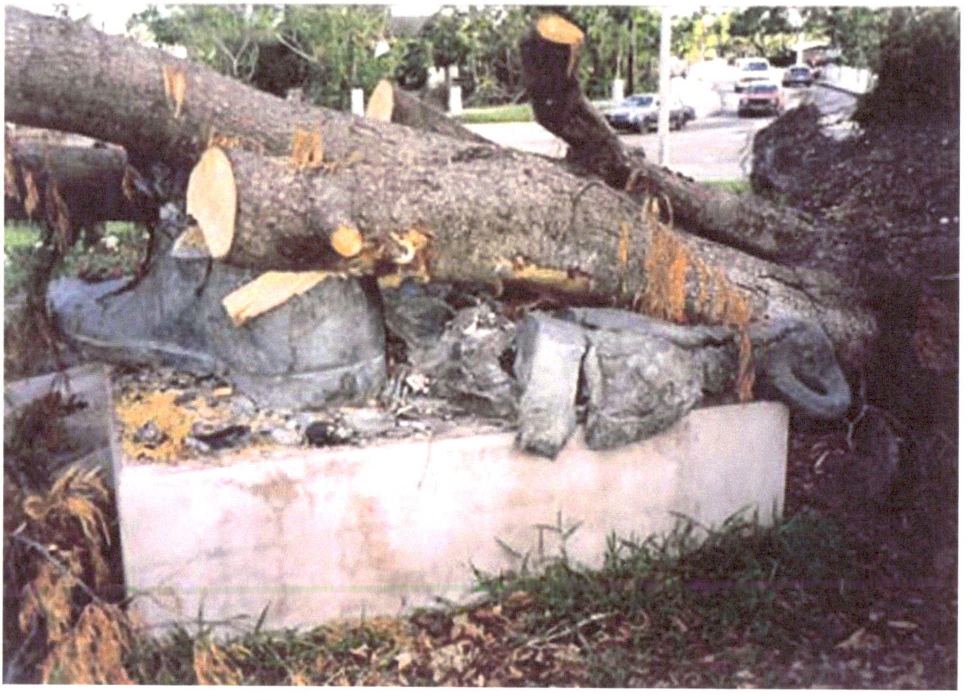

Figure 4 Damage caused by Hurricane Andrew in 1992 at the Pioneer Museum in Florida City.

full force of the hurricane, with 150 mph winds and rain. The roof was torn off and all the windows were blown out. Everything inside was wet. Adjacent to the museum was a small wooden train station, with track that probably went 600 miles to Louisiana. It was decided to completely evacuate the collection. I was equipped with a 'red phone' in order to contact Washington with my list of requirements – from the Smithsonian I arranged for a conservator to deal with clothing and the Library of Congress supplied a conservator who dealt with maps. There was an enormous amount of paper requiring attention.

Strange things happen when dealing with disaster. During my absence of about two weeks, the US Army had come in and encapsulated the house with plastic, making a kind of a hothouse for mold – there was orange mold on, under and in everything. We all know that in nature, when you see something orange, you shouldn't touch it or breathe it. We sampled the mold, bagged it and sent it to be analyzed and were told, 'Don't touch it, don't breathe it!' The mold specialists also informed us that the spores were not indigenous to the US –they were a gift from Hurricane Andrew.

We emptied out the house and wrapped everything up. All the paper objects were wrapped in butcher's freezer paper. The fabric and clothing were aired out on the porch and then lightly brushed to remove loose dust and spores and vacuumed. They were then wrapped in muslin and polyester marquisette and put into the freezer trailer. Every item in the house was removed.

The house then needed to be disinfected. There was a lot of discussion about this – the army wanted to spray everything with fungicides but I persuaded them that people would need to return to work in these spaces and that the residual toxic chemicals would continue to off-gas and be deleterious to human health. As we were 95 miles from the nearest hardware store, we obtained some high octane gasoline from the nearest gas station. Then we washed all the walls, the floors, the ceilings, the basement and the outside of the house. We left all the windows and doors open. After four days it was totally disinfected and habitable.

Valmeyer flood, Illinois, 1993

Sometimes people don't understand or accept when a disaster is inevitable. The Mississippi overflowed and flooded a little town in Illinois called Valmeyer in 1993. In fact, from 1880 until 1993 it had suffered ten floods, the majority 10–15 feet high. In 1993, a 25-foot flood swallowed the town. It was at that point that the citizens of the town finally decided to pack up and relocate the town on higher ground.

Deutsche Bank vault, New York City, 2001

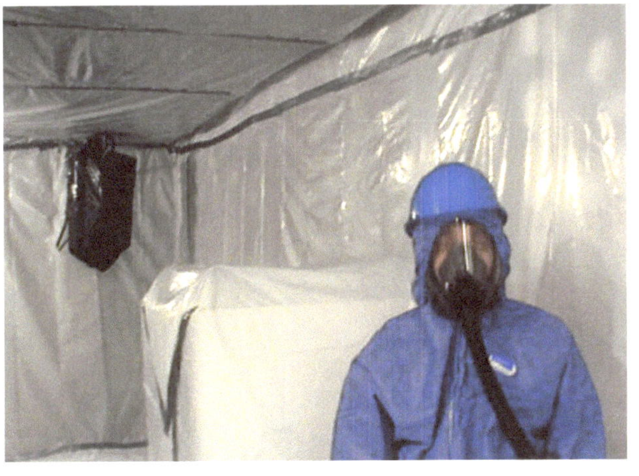

Figure 5 Ground Zero at the collapsed World Trade Center in New York (2003).

Two beams from the south tower of the collapsing World Trade Center destroyed floors four through nine in the adjacent building. The environmental controls failed and the two sub-basements filled with water. The bottom level contained all of the storage vaults. I was contacted by the environmental group overseeing the clean-up in February 2003. No one was allowed to go into the building until it was cleaned. The scope of my work was to set up all of the protocols necessary to dry out and recover as many of the documents in the vaults as possible.

Before I went to view the site, I had to take a three-day course in asbestos. My mask was fitted and tested to ensure a perfect seal. The entire area that I would be working in was contaminated so I had to wear three layers of Tyvek suits, Tyvek boots, and gloves.

My oxygen filter system lasted one hour and twenty-five minutes (Fig. 5). When I first visited the site there were 12 inches of water on the floor, the ceiling was dripping and the walls were covered with mold. I arranged for the water to be removed and the hallway, walls, and ceiling to be covered with four layers of 4mm thick polyethylene; each layer was sealed. The vaults had electric locks so a specialist was summoned to blow out the locks – an impressive spectacle I had only seen in the movies. The relative humidity in the vaults was 98%. Humidity and temperature monitoring devices were installed and a system set up with fans and dehumidifiers to bring down the humidity to about 75%. Since everything was contaminated, nothing could be removed from the site. A drum dryer was set up on site to dry the material. As a final step, a team of paralegals and lawyers viewed and approved all the documents that were recovered. From a seemingly lost cause, we managed a 95% success rate.

Conclusion

My time as a student in Florence was a major part of the foundation of my career as a conservator and one of the most rewarding experiences in the 40 years I have been practicing. In Florence, I was instructed in how to see. I learned that conservation was a very precise craft. I honed my skills in the following years with Bernard Rabin who firmly believed that conservation ranked as one of the highest crafts.

The experience of an American mud angel

Peter Mallory

November 1966 was a unique time both in my life and in that of this world. President Kennedy had been assassinated only three years earlier, and within two years there would be several more assassinations. I was a senior in college. I was involved in the civil rights movement. I was questioning the Vietnam War. I was an angry and frustrated and disappointed person.

But I was also a student in art history at the University of Pennsylvania, and Dr Frederick Hartt was my professor of Renaissance art history. When he came into class on Monday morning, November 5th, 1966, he had tears in his eyes and told us that he was leaving immediately for Florence and would be back in a week. He came back the following Monday, again in tears, with slides and a first-hand account of the devastation and of all the students and volunteers who had just appeared to help. At the end of the session I told him: 'Dr Hartt, I'm going to Florence. As soon as exams are over in December, I'm going.'

And I made it happen. I wrote a series of articles for my hometown Indianapolis newspaper that was republished around the country. During my stay, I did sculpture restoration throughout the city (using the Hempel method) and when I think back to my times in Florence they were among the happiest of my life. The decision to go to Florence was a no-brainer – back in the '60s we wanted to change the world, but my world at home wasn't cooperating. In Florence, I *could* make a difference. Florence was one of the few times when I felt I was a real *participant*. Of course, what happened was a terrible tragedy, but I didn't choose the tragedy – I just chose to act upon it.

When I was preparing for my trip, Dr Hartt told me, 'You cannot stay in Florence. There is no power, no light, no water. You must stay in Fiesole. That is the only place where the *pensioni* are still open. You must stay at Pensione Bencistá.' That was where he had stayed on his most recent trip, and that was where he and Bates Lowry, Fred Licht and all the people who founded the Committee to Rescue Italian Art had resided. They would go to I Tatti, talk, make decisions and come 'home' to Bencistá. So I sent a telegram to Pensione Bencistá, and about the 18th or 19th of December, I flew to Milan. I did not speak Italian, but somehow I made it onto the train to Florence, got off at the *stazione*, found the No. 7 bus, and reached the piazza in Fiesole.

I looked around and asked for directions. Finally, through some pointing and a little bit of French, I asked for directions to Pensione Bencistá, and they pointed me down the hill one kilometer. Well, in those days before the invention of rolling bags, I had to carry my suitcase all the way. It was 8.00 pm and pitch black – I had been up for 30 hours. I finally found Pensione Bencistá, walked down the drive and knocked on the door. The owner, Signor Simone, came to the door and I announced myself only to be told 'We're closed.' Luckily, unlike the Christmas story of 2000 years ago when another family went to the inn and was turned away, Signor Simone paused a moment and then said: 'Well, come in', and so I became a member of their family. I ate at the family dinner table. Signor Simone's father (the grandfather of the current generation) had a little BMW motorbike, and I rode down with him to Florence every day and came back with him every evening. I have returned to Pensione Bencistá several times in the 40 years since then – the family that took me in then are still my Italian family.

But their kindness was typical of the things that were going on throughout the city in those days. All the conservators in the Palazzo Davanzati and all the book people who were laboring in the Biblioteca Nazionale Centrale di Firenze were seeing one frustrating and tough reality. But what was most important for the city in the days before Christmas was that the pope was coming on Christmas Eve and would say midnight mass in the Duomo. So in preparation, the entire area from Palazzo Pitti to San Marco, from Ognissanti to Santa Croce, had pretty much been cleaned.

The weather was gorgeous. The air was still. Every day the sky was an impossible blue. The lights were back on, and there was glass back in most of the shop windows, even if there was hardly anything to display. I remember a shoe shop, a very fashionable one, with just one pair of shoes; but there was a spotlight on them, and they were open for business and celebrating Christmas.

After about three or four days, the Simones had had enough of their American child and found me a *pensione* that had just reopened in town on Via Tornabuoni, so I was delivered down at the end of the day. I was so excited to be actually in Florence that I put on my running shoes, and I ran for about two hours. I ran down every street because I wanted to learn the city completely. I had my little map, and I would run from one end to the other, up and down – I must have run about 15 miles. I was very sore the next day, but I was so elated.

On Christmas Eve I went to the Duomo for the midnight service, and the whole city turned out. They were so proud that the pope had come to support them. The crowd was so big that I could only occasionally see Pope Paul directly. I was 100 meters away at the far end of the nave, but there was a television set there, so that's how I watched the service. That was just a wonderful night.

It is important to remember when we hear about all the terrible things that happened in 1966 that there also was hope. There was another *Rinascimento* in Florence after the flood, and the citizens of Florence and all the volunteers deserve a great deal of credit.

When I arrived, most of the other mud angels (*angeli del fango*) had gone home for Christmas, so I found plenty of opportunities to work. I worked with the UNESCO film crew and then at the Museo dell'Opera del Duomo for another several days after Christmas.

I returned to Florence the following summer under the auspices of CRIA, working from 9.00 am to noon and then from 3.00 to 6.00 pm, something like that. As I didn't need the full siesta, my friend and co-worker from Harvard, a fellow named Steve Herbert, and I had some time on our hands each day. Our landlady that summer (who we called Aunt Maddie) was acquainted with the head of the Guild of the Misericordia, the ambulance drivers for the city. Their headquarters with all the ambulances out in front is right next to the Duomo (to the right of the façade as you look at it). They have a lovely chapel, very much in the style of the Old Sacristy, full of *pietra serena* columns and pilasters as well as a beautiful della Robbia Madonna, but as it was built only relatively recently (in the nineteenth century), it was way down on the Soprintendenza's priority list, and the way things were going, it wouldn't be cleaned for a decade or more.

Steve and I decided to take on the job ourselves. At noon each day, we would wolf down a calzone at our favorite pizzeria and then rush off on Steve's Lambretta to the materials dump to get a pail of talc and another of solvent, and we spent the entire summer as freelance art conservators cleaning the chapel of the Guild of the Misericordia.

When we first saw it, there was what looked like a bathtub ring of 'nafta' or heating oil in the *pietra serena* all around the entire chapel. *Pietra serena* is quite porous and it would take many 'poultices' over many weeks to remove the nafta from the walls and from the altar. The Misericordia headquarters is not a public place, but we became so much a part of the normal routine of the guild that we might as well have been invisible to the members. We would just go in and out, day after day, and soon nobody seemed to notice us. Neither of us spoke Italian at the time and none of them spoke English.

At about 1.00 pm every day, the janitor would come in, go to the altar, remove all the candles that had been lit during the morning so that in the afternoon people could buy new candles and light them as they prayed. He would extinguish them and put them in a box behind the altar. It was pretty routine, and I had seen it done in all the churches that I worked at – it happened in the Badia and in Santa Croce. After about two months and on our very last day, the janitor came in, did his thing and left. We took little notice. We continuing working, wrapped up in finishing our work, and suddenly Steve turned to me and asked: 'Do you smell something?' Apparently, one of the candles in the box had not been completely extinguished, and after half an hour or an hour in this wooden box filled with hundreds of other candles, it finally caught all of the tallow on fire, and the beautiful brocaded cloth hanging behind the altar was now being consumed by 8 feet of flames, reaching up to the ceiling.

Neither of us knew the Italian word for fire or emergency or help so Steve ran out into the main hall yelling, 'Attenzione! Attenzione!' pointing and dragging people to try to get them to come with us to the chapel. They finally got the message and took care of the fire, but we were so embarrassed, afraid that they might suspect that we had caused the fire ... and we could not *explain* our innocence. Luckily, within seconds we were again invisible as they took care of cleaning up so we said nothing, and when our job was completed and all the nafta had finally been removed, we snuck out. The next day we flew home, wondering if they were cursing those foolish American boys! However, six months later, apparently after Aunt Maddie gave them our addresses, we both received lovely thank you notes from the Guild of the Misericordia.

When I returned a few years ago with my wife, the janitor remembered me. Luckily, he didn't remember the fire – he just remembered me. We were led upstairs and treated like royalty.

Being an *angelo del fango* was a wonderful experience – I only have happy memories of Florence and the flood.

Recovery of the panel paintings of Florence

Cristina Acidini

At the time of the flood in November 1966, Florence became an international laboratory of conservation experts who converged on the city from all over Europe and the United States – from every part of the world interested in Western art. Their combined efforts and the inevitable exchange of information and ideas transformed the celebrated Florentine tradition of restoration, raising it from the provincial level to one of international respect and importance. It was the flood that gave impetus to the creation of the modern Opificio delle Pietre Dure with its various restoration workshops. The Opificio had been so named since the Lorraine period but had existed under the Medici; it was already an organised group of workshops by 1588 or even earlier. Immediately following the flood, at the time of the emergency, the restorers of the Opificio, who concentrated on sculpture and the decorative arts, worked side by side with the restorers from the Gabinetto dei Restauri della Soprintendenza alle Gallerie, founded in 1932 by Ugo Procacci, who specialised in restoring paintings on panel and canvas.

Ugo Procacci, superintendent at the time, supervised the united efforts of these two specialist bodies, also incorporating the operations carried out by experts from all around the world. He was assisted by the young Umberto Baldini and the even younger Paolo Dal Poggetto, who, alongside Luciano Berti, Luisa Becherucci and so many others, worked for the Soprintendenza in those years. They combined forces with the Opificio restorers to set in motion the operation to recover Florence's rich heritage.

Some of the images shown below serve to illustrate the methods and developments at the time with regard to one of the art forms most severely affected by the disaster: panel painting. Panel painting is a form of artistic expression particularly widespread in Florence and Tuscany. Well before the flood, Ugo Procacci and his assistants had devoted considerable attention to panel painting and problems related to restoration and many innovative solutions were developed in this field.

The map of Florence (Fig. 1) shows those streets (marked in brown) most affected by the muddy water and gives some idea of the extent to which the centre of the Florence was badly hit, both to the east around Santa Croce, in the centre around the Duomo and Santa Maria Novella and in some parts of the Oltrarno. In Figure 2 we can see the Baptistery doors standing in the filthy oily water: the east door by Ghiberti was pushed open and banged about on its hinges. Five of the panels fell out into the waters below. Figure 3 (from a magazine) might not record an actual event but it is a scene that might well have taken place. It shows Ugo Procacci, standing in the muddy water of the Uffizi storerooms, lifting a picture out of danger and handing it to Umberto Baldini, who is shown moving it to safety.

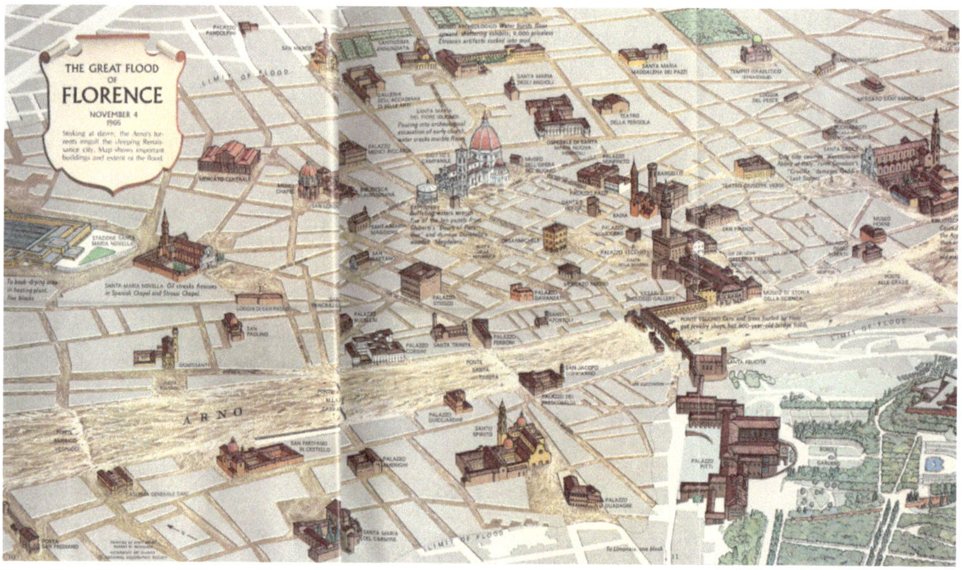

Figure 1 Map of Florence showing those streets (marked in brown) most affected by the muddy water.

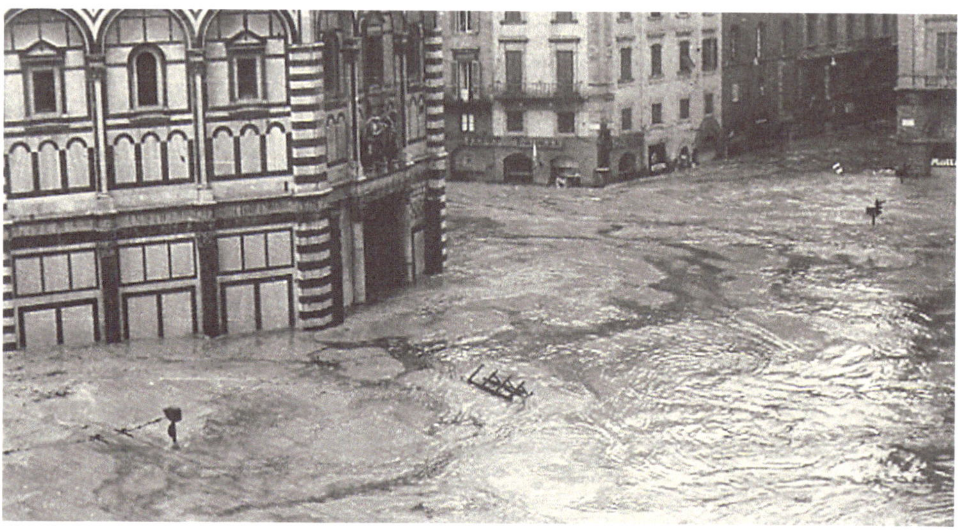

Figure 2 The Baptistery doors standing in the filthy oily water.

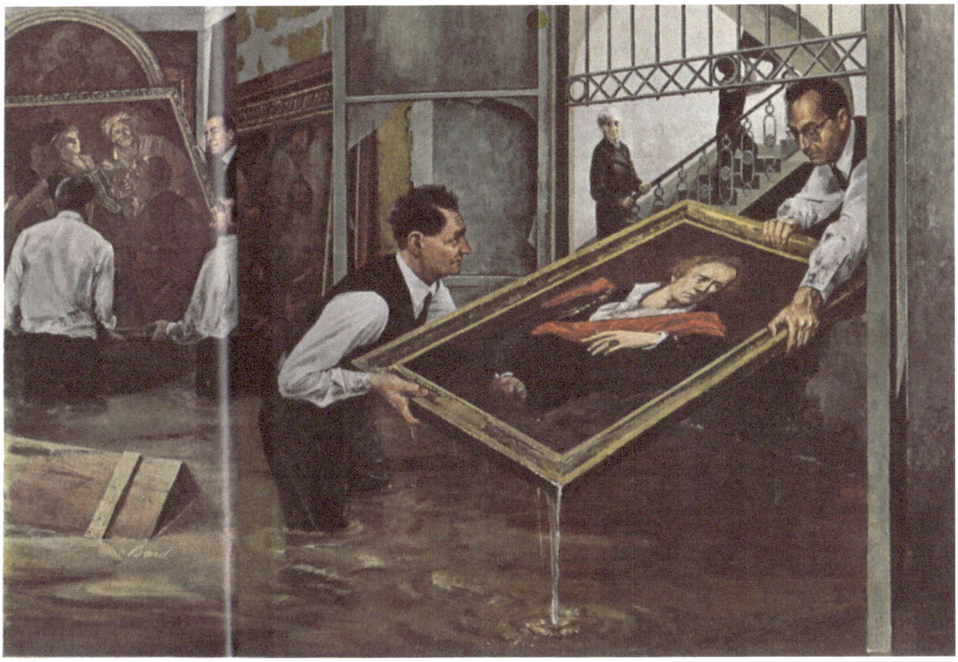

Figure 3 This illustration (from a magazine) shows Ugo Procacci, standing in the muddy water of the Uffizi storerooms, lifting a picture out of danger and handing it to Umberto Baldini.

The Vasari Corridor in the Ponte Vecchio, then, as now, was full of paintings. On the night of 3 November, Luisa Becherucci, director of the Uffizi, battled to save what she could because of fears that the bridge might collapse. In the event, the bridge remained firm but Becherucci acted under instructions from Procacci, who thought any risk was worth taking to save the pictures. Although the bridge did not collapse, the storerooms of the Vecchia Posta and others in the Vasari complex were all flooded. Figure 4 illustrates one of the most important centres set up to cope with the disaster – the sculpture workshop at the Bargello. Figure 5 shows the large crucifix painted by Cimabue with Baldini (on the right) – his efforts were of fundamental importance in the months and years following the flood. Figure 6 illustrates the cross being worked on the Fortezza da Basso laboratory.

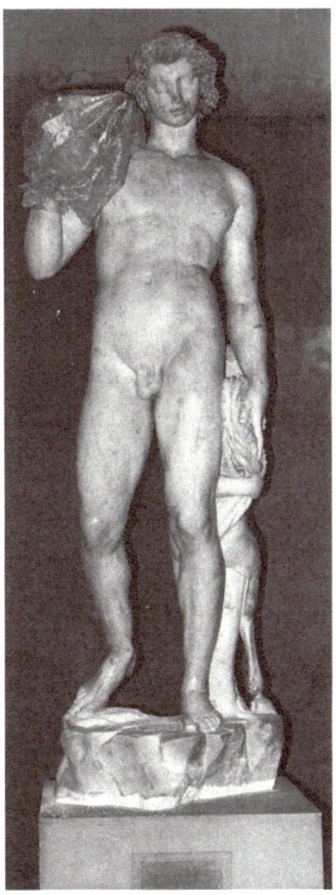

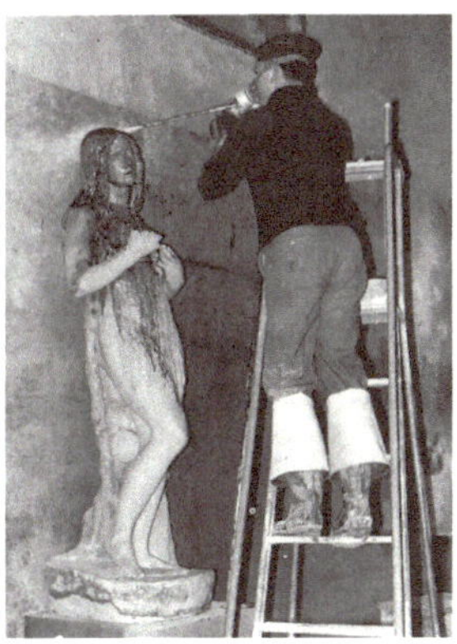

Figure 4 The sculpture workshop at the Bargello.

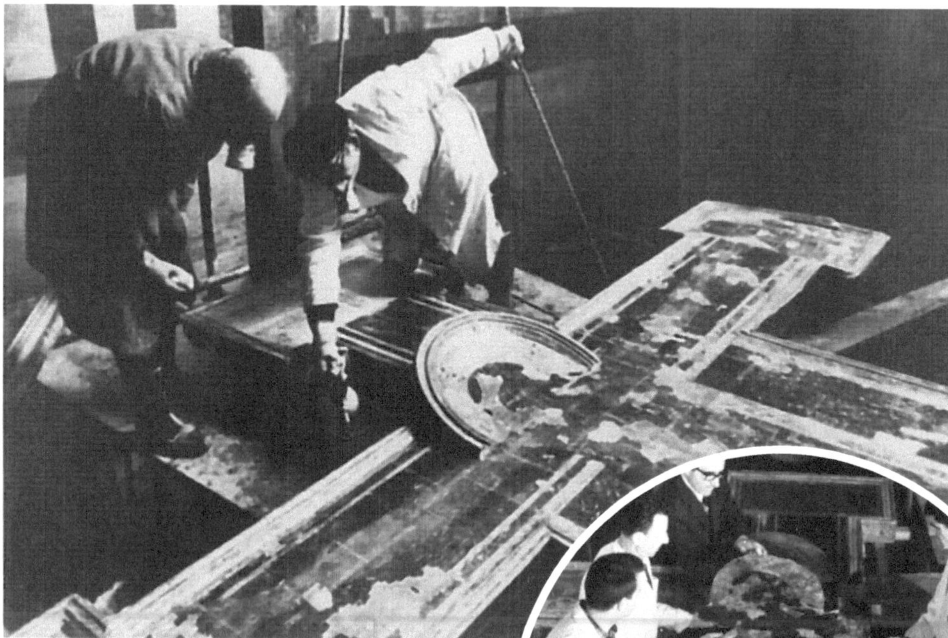

Figure 5 The large crucifix painted by Cimabue: Baldini can be seen on the right in the photo.

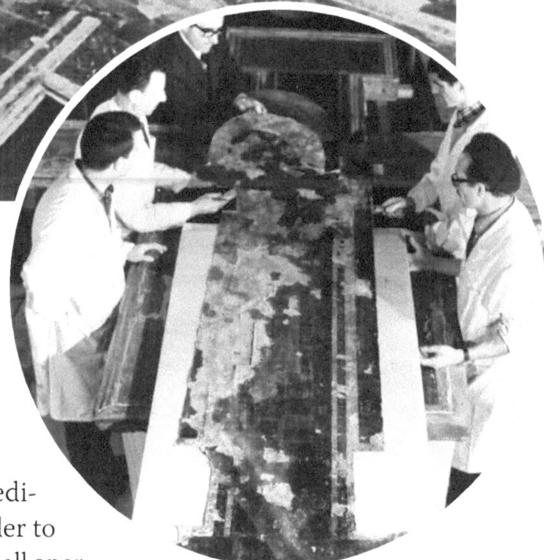

Figure 6 This photo shows the cross by Cimabue being worked on in the Fortezza da Basso laboratory.

Some decisions were put into effect immediately to create a controlled environment in order to monitor the process of wood drying and to keep all operations and works of art in Florence. This was something of a strategic choice, but also a courageous one for a city that was just emerging from disaster. It reflected confidence in Florence's ability to cope: only the historic arms and armour were removed to the Rüstkammer in Vienna. On the other hand, material and experts were brought in from all around the world and a new laboratory was created at the Fortezza del Basso. Figure 7 is an early picture of a work arriving at the Limonaia in the Boboli Gardens with Japanese paper glued to it in order to keep the paint – which had already begun to swell and flake off the surface – in place.

The pope is shown in Figure 8 on Christmas night – clearly moved – before the cross by Cimabue. The Limonaia workshops are shown in Figure 9 and the photograph of the Fortezza da Basso laboratory (Fig. 10) shows it very like it is today although it is now more technologically advanced following the renovations in 1999. The greatest damage to panel paintings was that caused by the accumulation of mud on the paint and the impact of the water that destabilised the natural cohesion created by the painter between

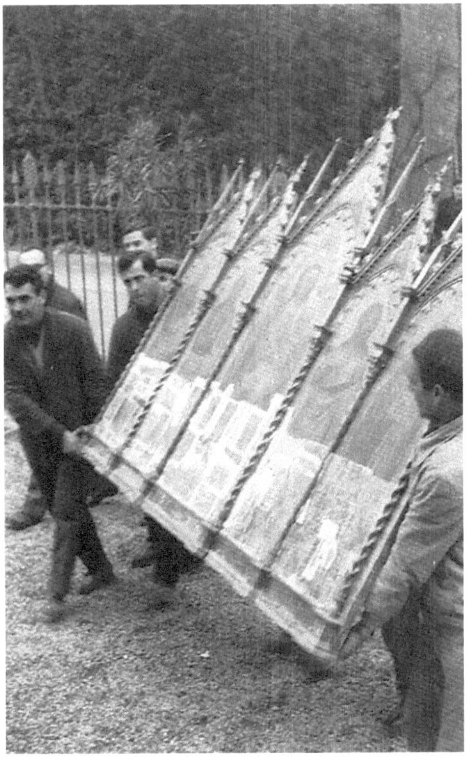

Figure 7 An early picture of a work arriving at the Limonaia in the Boboli Gardens with Japanese paper glued to it in order to keep the paint in place.

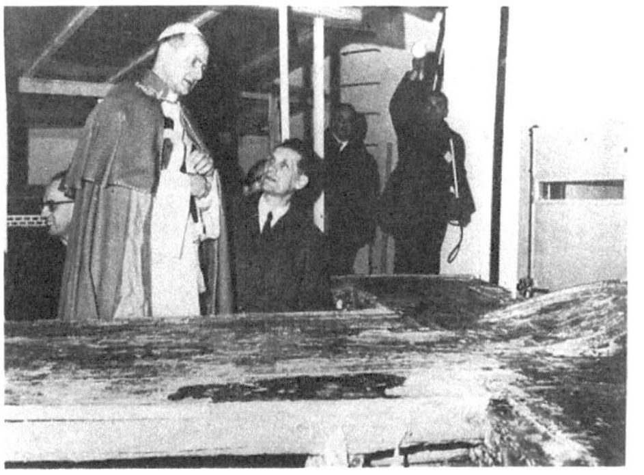

Figure 8 The pope on Christmas night before the cross by Cimabue.

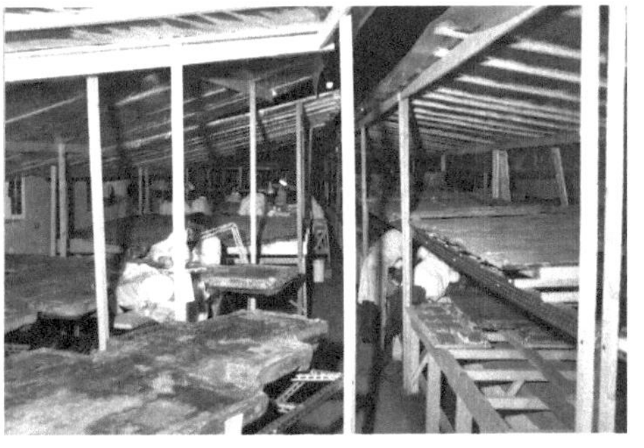

Figure 9 The Limonaia workshops.

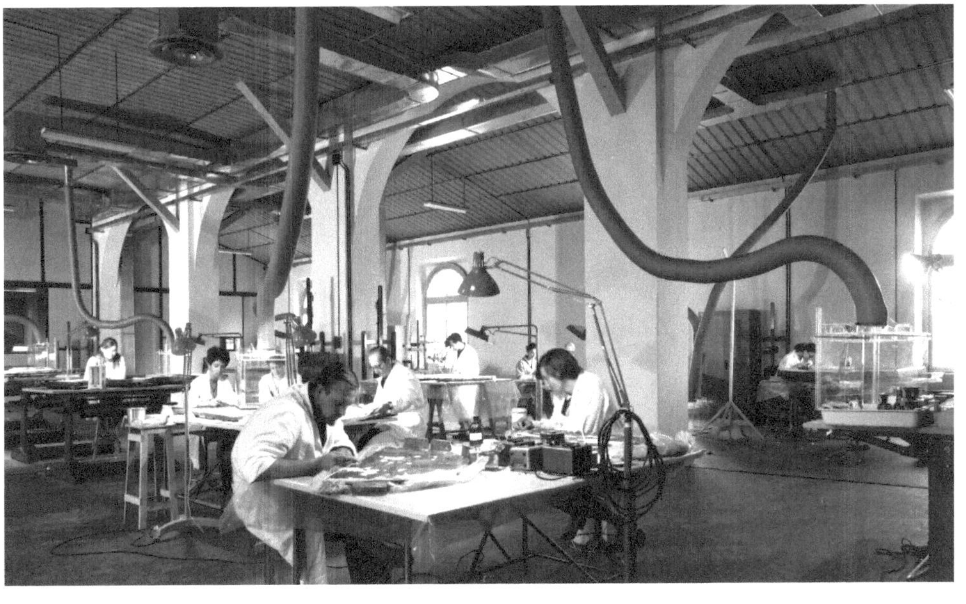

Figure 10 The Fortezza da Basso laboratory.

the paint layers, the preparatory layers and the wooden supports. The lifting of these layers and the expansion and contraction of the wooden support caused buckling and cracks in the paint. Wood of course, once it has expanded with moisture, often shrinks to less than its original size as it dries. The most celebrated work to have suffered most from this process was the Cimabue crucifix, which soon came to symbolise the tragedy. Almost immediately, through Baldini's devotion, it also became the symbol of recovery. A triage or screening system was adopted in order to give priority to those works most in need of immediate attention and to help classify their various problems. In the pivotal *Florence Restored,* the group of restorers photographed here with Procacci (Fig. 11) provided the inspiration and driving force behind the 1972 exhibition, with its survey of 40 years of restoration history, though understandably a large proportion of the works on display were victims of the flood.

Among these was the Cimabue cross (Fig. 12). and beside it is the table of values from the Fortezza archives designed to show the period of time over which wood contracts. A year was needed for the wooden support to return to its original size during which time it had to be kept under careful control so that the process did not accelerate and cause further damage. It was then necessary to detach the paint, and at this stage we were helped greatly by the skill of Cimabue. He had made sure his assistants covered the support with a single piece of cloth which, therefore, was relatively easy to detach.

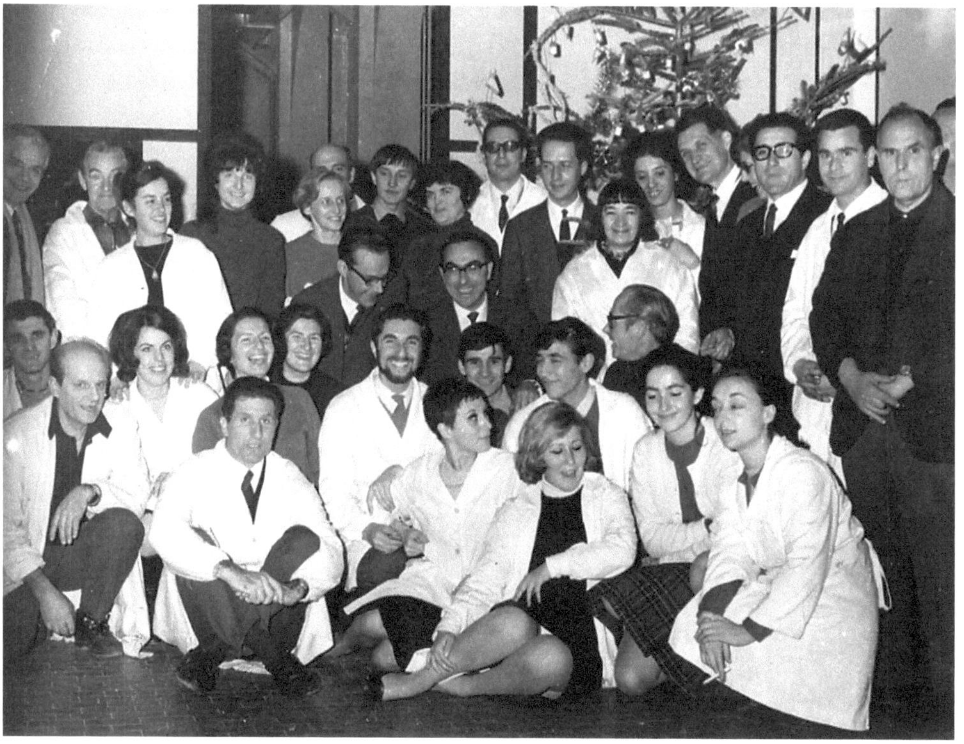

Figure 11 The group of restorers photographed here with Procacci provided the inspiration and driving force behind the 1972 exhibition *Florence Restored.*

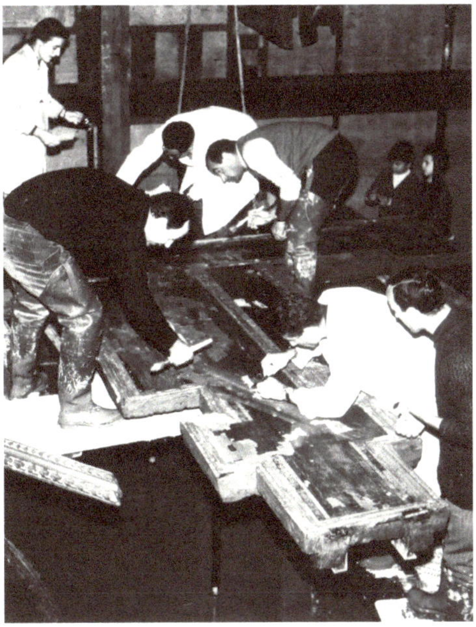

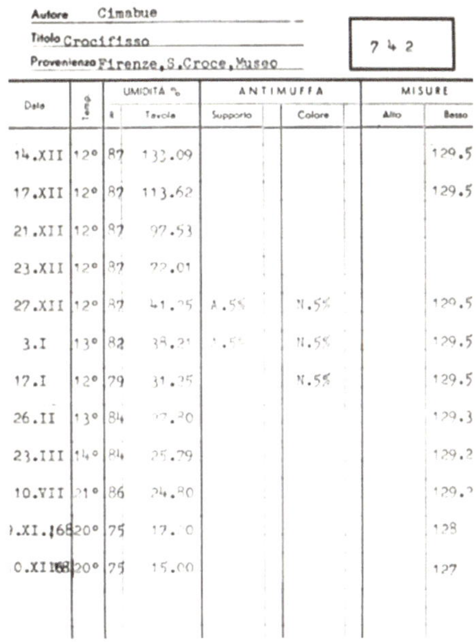

Autore Cimabue

Titolo Crocifisso

Provenienza Firenze, S.Croce, Museo

7 4 2

Data	Temp.	UMIDITA %		ANTIMUFFA		MISURE	
		%	Tavole	Supporto	Colore	Alto	Basso
14.XII	12°	87	133.09				129.5
17.XII	12°	87	113.62				129.5
21.XII	12°	87	97.53				
23.XII	12°	87	72.01				
27.XII	12°	87	41.25	A.5%	N.5%		129.5
3.I	13°	82	38.21	A.5%	N.5%		129.5
17.I	12°	79	31.25		N.5%		129.5
26.II	13°	84	27.20				129.3
23.III	14°	84	25.79				129.2
10.VII	21°	86	24.80				129.[illegible]
[illegible].XI.168	20°	75	17.[illegible]0				128
0.XII68	20°	75	15.00				127

Figure 12 The Cimabue cross and the table of values from the Fortezza archives designed to show the period of time over which wood contracts.

Having been completely removed and then restored, it was reattached with an intermediate layer of synthetic material. Restorers often liken panel paintings with their many layers to sandwiches and this one was enriched with an additional filling. The final result was achieved using an experimental approach to filling in the pictorial losses, adopting Baldini's method of chromatic abstraction and chromatic selection, treating the lacunae in the painted surface with a series of dynamic and overlapping brushstrokes picking up the surrounding colours.

Lippo di Benivieni was equally ingenious in his cross (Fig. 13) where a number of different pieces of cloth supported the painted surface. These had to be detached separately and then recomposed. Figure 14 illustrates a piece lifted from the wooden support, clearly showing the 'sandwich effect'.

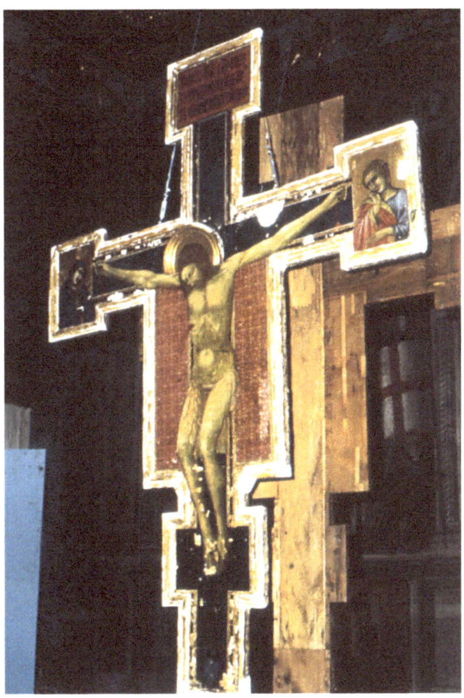

Figure 13 The cross by Lippo di Benivieni.

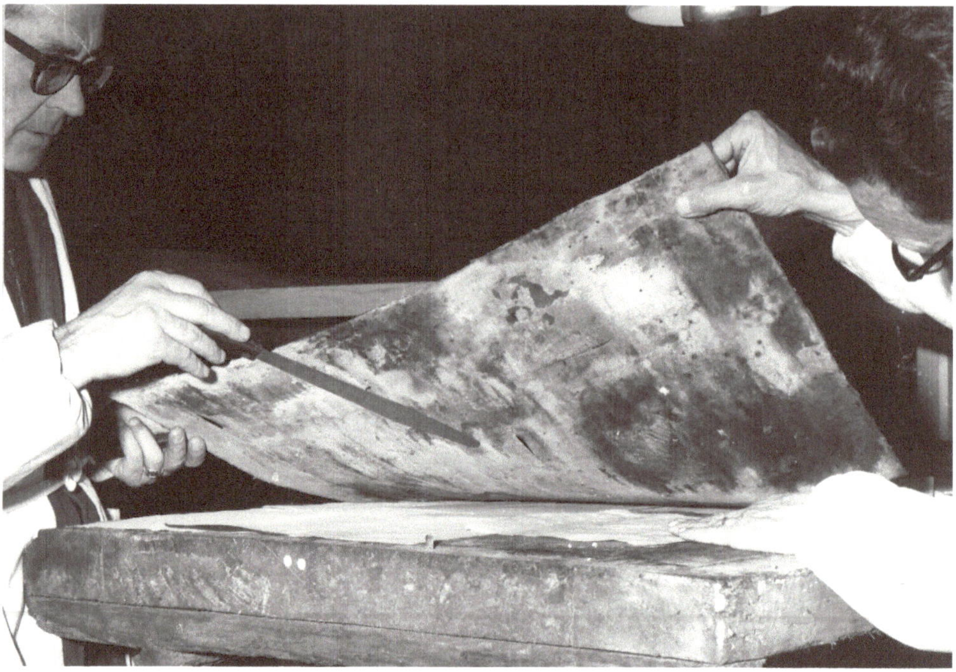

Figure 14 The cross by Lippo di Benivieni: a piece lifted from the wooden support, clearly showing the 'sandwich effect'.

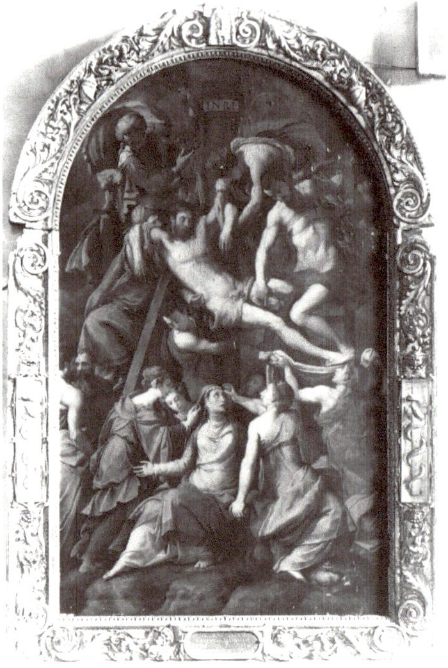

Figure 15 The large altarpiece by Francesco Salviati: the painting before restoration.

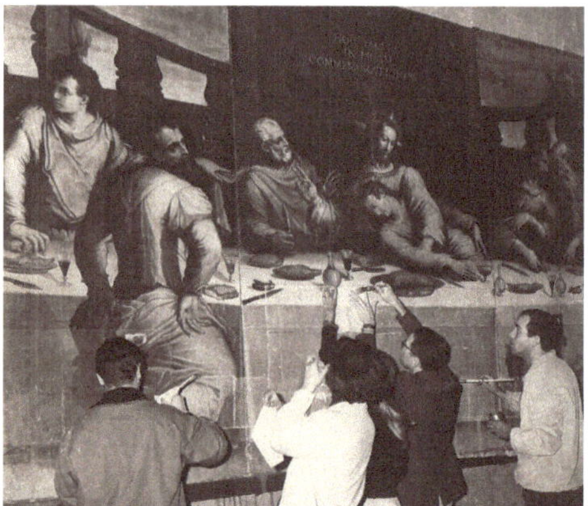

Figure 16 The *Last Supper* by Giorgio Vasari: an operation carried out on the spot at the time of the flood: the *velinatura* or application of Japanese paper.

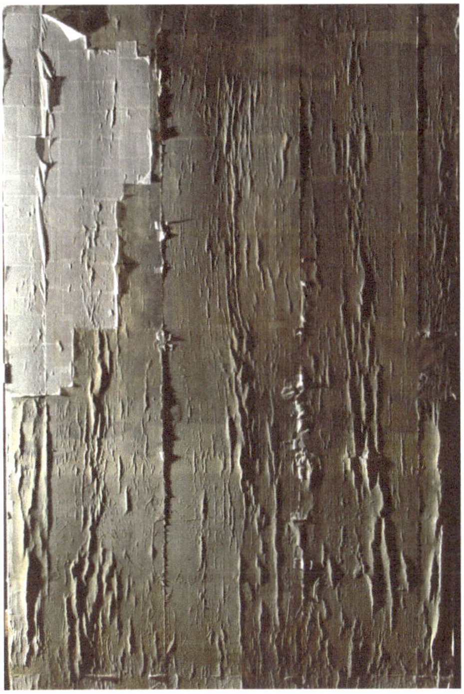

Figure 17 The *Last Supper* by Giorgio Vasari: a detail of the painting more than 40 years after the flood.

The large altarpieces by Francesco Salviati, as well as those of Bronzino and Allori, were in a pitiful condition, as were their frames. The paint was cracked as is the way with panel paintings. Figure 15 shows the painting before restoration. Figure 16 illustrates an operation carried out on the spot at the time of the flood: the *velinatura* or application of Japanese paper to the *Last Supper* by Giorgio Vasari. A detail of Vasari's painting today, more than 40 years after the flood, can be seen in Figure 17. But new funds made available to the Soprintendenza per il Patrimonio Artistico under Bruno Santi hold out great hope of being able to recover much of what had appeared lost. X-radiographs and reflectographs have shown the condition of the painting underneath the flood damage, but initial work has also clearly revealed the head of St Peter, the one turned abruptly towards Christ.

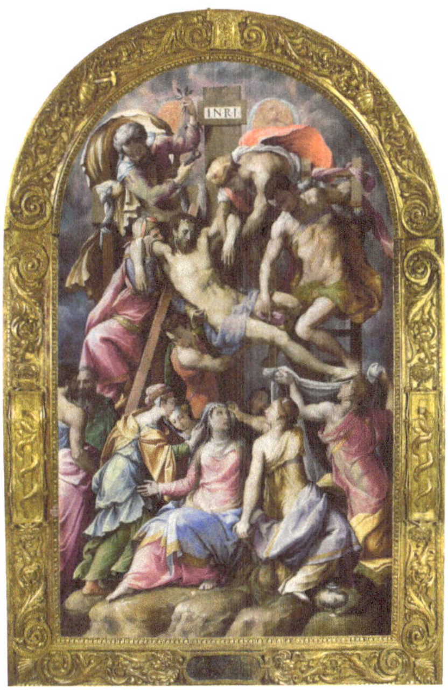

Figure 18 The *Deposition* by Francesco Salviati shown in its original monumental frame, now beautifully restored.

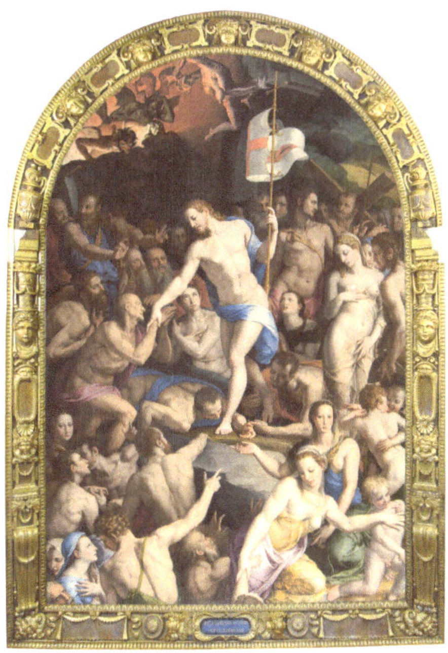

Figure 19 Bronzino's *Descent of Christ into Limbo*, also with a very fine frame.

Other works restored at the Fortezza da Basso by the Opificio delle Pietre Dure can be seen in the Cenacolo and Santa Croce Museum where they are displayed together with other restored works from the museum. Works by masters, perhaps not of the level of Cimabue, but which have received attention and laborious devotion from our restorers under Marco Ciatti and have now finally returned home include those by Giovanni del Biondo, Lorenzo Monaco, Rossello di Jacopo and Domenico Nichelino. The *Deposition* by Francesco Salviati (shown in Fig. 18 in its original monumental frame, also now beautifully restored) and Bronzino's *Descent of Christ into Limbo* (also with a very fine frame, Fig. 19) have returned with the other works to the Museum of Santa Croce as part of a temporary exhibition but they will eventually join the permanent display. This is a tremendous advance and almost the conclusive stage, now only awaiting the restoration of the *Last Supper* by Vasari, in the recovery of an artistic heritage so severely damaged by the flood – the panel paintings of Florence.

Editor's note

Translation by Heather Roberts.

The United States Heritage Emergency National Task Force: disaster recovery now and in the future

Debra Hess Norris,[1] Jane Long and Mary Rogers

In 1995, the Heritage Emergency National Task Force was formed to help libraries, museums, historical societies, and historic sites better protect their collections and buildings from natural disasters and other emergencies. A partnership of 41 US federal agencies and national service organizations, the Task Force is co-sponsored by the national non-profit Heritage Preservation and the US Federal Emergency Management Agency (FEMA). Its work is sustained by public and private projects' grants and contributions. The mission of the Task Force is to:

- Help cultural heritage institutions and sites better prepare for emergencies.
- Encourage the incorporation of cultural and historic assets into disaster planning at all levels of government.
- Facilitate a more effective and coordinated response to all kinds of emergencies.
- Assist the public in recovering damaged heirlooms.

In the past decade, the Task Force has developed planning tools, promoted safe salvage procedures, advanced practical recovery strategies for cultural institutions and the public, and fostered collaborative response networks. Significant outcomes led by Heritage Preservation and described in detail below include the *Emergency Response and Salvage Wheel,* which has been translated into six languages; the new *Field Guide to Emergency Response,* a handbook and DVD designed to help staff and volunteers respond safely to disasters; Alliance for Response, city- or county-wide forums that build partnerships between cultural heritage stewards and first responders; and *Cataclysm and Challenge*, the first comprehensive overview of damage to cultural heritage from the September 11th attacks.

Emergency Response and Salvage Wheel

The *Emergency Response and Salvage Wheel*, originally published in 1995 and updated in 2005, provides emergency response steps and wet salvage information in an easy-to-use slide-chart format. Thousands of copies are in use in museums, libraries, and archives worldwide.

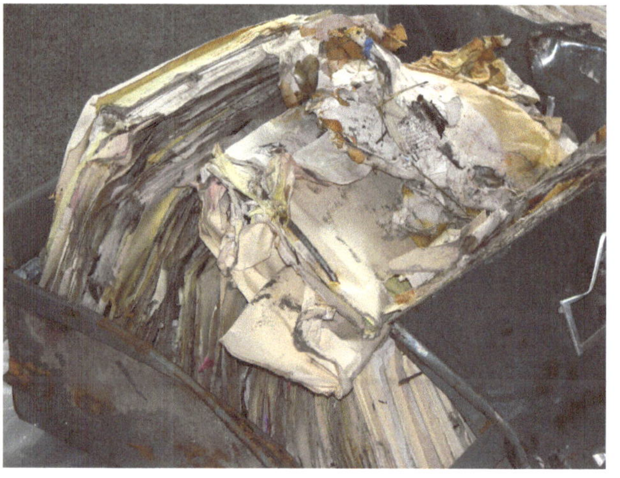

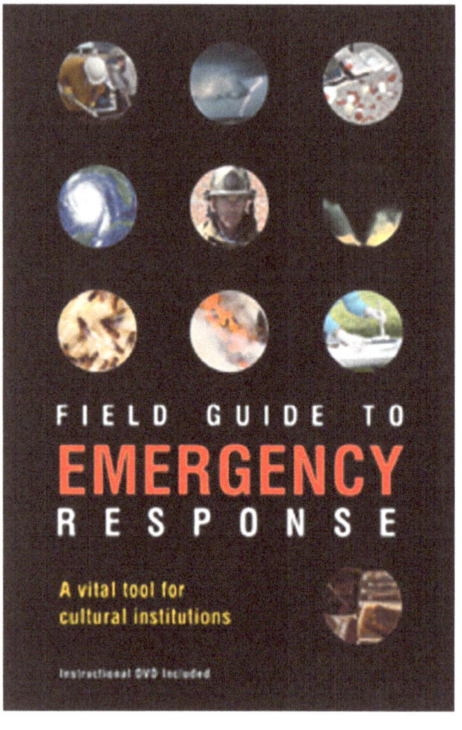

Field Guide to Emergency Response

The award-winning *Field Guide to Emergency Response*, published in 2006, explains clearly and simply the steps to take in the first few hours of a disaster, enabling even those with no prior training to save collections without endangering themselves. In addition to an easy-to-follow handbook, a companion DVD illustrates salvage techniques for typical problems like mud, mold, and soot. The DVD can be used on-site in a laptop or vehicle player, as well as for preparedness training. Also included are information panels that can be customized before disaster strikes and helpful checklists on supplies, vendors, and salvage priorities.

Alliance for Response

Alliance for Response, initiated in 2003, facilitates partnerships between cultural heritage professionals and local emergency responders, beginning with a one-day forum. The forums provide an opportunity for cultural heritage staff of all levels – from executive directors to curators to conservators – to exchange important information with and get to know their local firefighters, police, and emergency managers in order to ensure the best possible outcomes in emergencies. The forums often result in local networks and new cooperative projects. The Alliance for Response initiative has reached more than 500 institutions and agencies in 15 communities.

Cataclysm and Challenge

Cataclysm and Challenge, published in 2002, offered the first comprehensive study of the damage sustained by cultural heritage on September 11, 2001, and how museums, libraries, and archives in Lower Manhattan were affected. It also presented an overview of the impact on the Pentagon library and military art collections. Key recommendations from the report called for increased staff training and for current collections inventories. The study also urged more effective communications between the emergency management and cultural property fields. The recommendations are designed to address any type of emergency and can be used by collecting institutions throughout the country.

Other initiatives by Task Force members include MayDay, designed by the Society of American Archivists to encourage readiness, and the 'Train the Trainers' initiative, developed by the American Institute for Conservation of Artistic and Historic Works and designed to inform conservation professionals about safe and efficient salvage procedures for library and photographic materials, works of art on paper, paintings, textiles, decorative arts, ethnographic objects, and electronic media.

Heritage Health Index

A recent comprehensive study of the state of preservation of America's collections – the *Heritage Health Index* – found that 80% of cultural institutions in the United States do

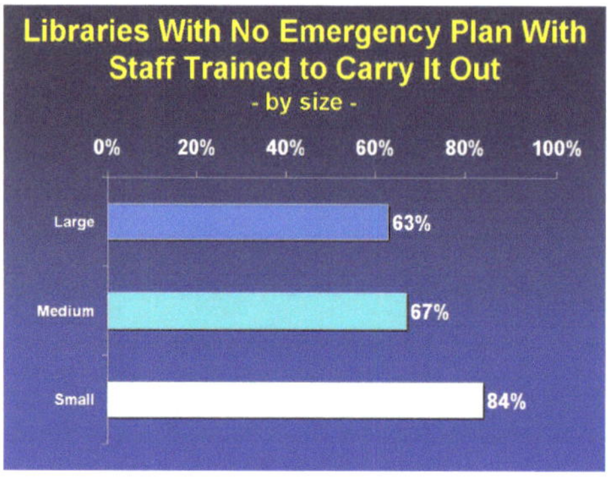

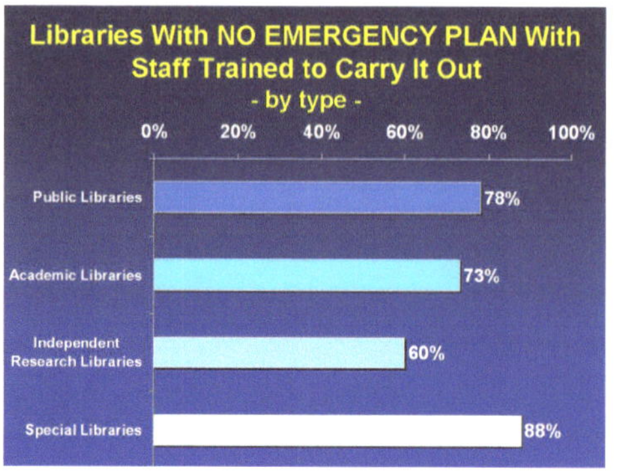

not have a disaster plan that includes collections and the staff trained to carry it out. Libraries (78%), historical societies (92%), and museums (78%) are especially unprepared. The data are also not encouraging when viewed by size of institution; 60% of large institutions in the United States have no emergency plan with trained staff. Heritage Preservation is pleased to be working with the Institute for Museum and Library Services on a pilot grant program aimed at providing small- to mid-sized cultural institutions with grant funds for emergency consultation and planning that will result in a customized disaster plan. New resources, such as *dPlan,* a free, on-line disaster planning tool created by the Northeast Document Conservation Center, will simplify this process nationally.

Graduate art conservation education and Hurricane Katrina response

In recent years, graduate programs in art conservation have also significantly strengthened course work in emergency planning and response, incorporating lectures on mold mitigation, disaster planning, and on-site drills focused on response and recovery. During the past two years, art conservation faculty and graduate students led by the University of Delaware and Winterthur Museum & Country Estate in Wilmington, DE – and including New York University, Buffalo State College, and the University of Texas at Austin – have provided collection recovery and preservation expertise to salvage collections on the Mississippi Gulf Coast devastated by Katrina. This national initiative in partnership with the Mississippi Department of Archives and History in Jackson, MS, has focused on Biloxi, MS, with special attention being given to two devastated institutions: the Ohr-O'Keefe Museum and Beauvoir, the Jefferson Davis Home and Presidential Library.

Primary goals included providing these and other cultural institutions with sustained assistance *and* future conservation professionals with real-life emergency response training and education. This challenging and ultimately rewarding work was funded with significant in-kind support as well as a $400,000 grant from the Andrew W. Mellon Foundation.

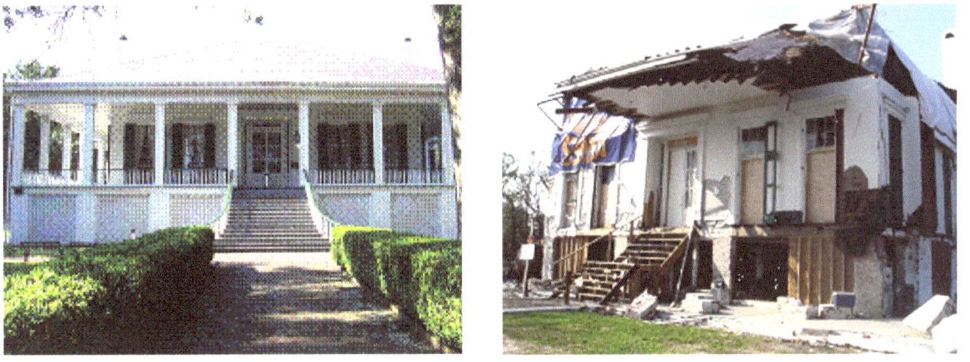

Figure 1 Beauvoir, home of Jefferson Davis, before and after Hurricane Katrina.

This initiative provided assessment and stabilization treatment for thousands of artifacts. Thirteen graduate and undergraduate students from conservation programs nationally participated in eight-week on-site summer programs. All students gained considerable practical skills and developed innovative conservation treatment and preventive care solutions for collection objects based on the extent of damage, curatorial priority, and conservation equipment and supply limitations. Discipline-specific teams of conservators, students, and/or recent graduates also visited the Gulf Coast for sequential four-day periods to offer a series of six workshops to 42 affected cultural institutions. Half-day workshops for the public were also offered. A limited number of damaged artifacts were packed and shipped for examination and treatment by advanced graduate students; these treatment projects offered our students challenging treatment experiences in many ways similar to those encountered more than 40 years ago in Florence – the consolidation of crumbled surfaces, the reduction of severely blanched coatings, mold removal, and extensive loss compensation.

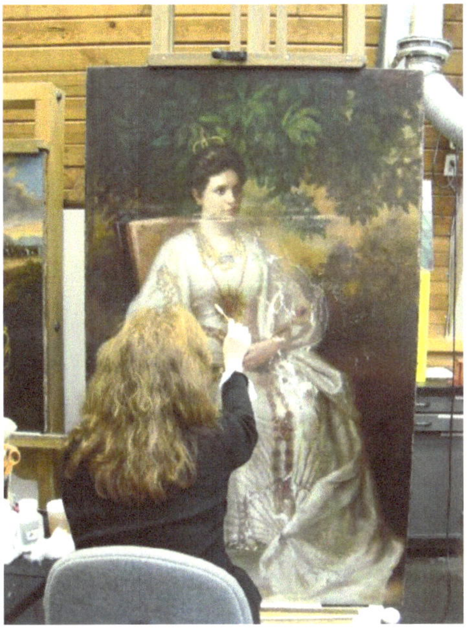

Figure 2 Amber Kerr-Alison, Winterthur/University of Delaware graduate student, treating a portrait of Winnie Davis from Beauvoir damaged by Hurricane Katrina.

These emergency response and educational activities benefited Beauvoir and the Ohr-O'Keefe Museum in substantive and significant ways, and provided expert assistance to other Gulf Coast cultural institutions in need. This partnership has served as a national model for emergency response and provided everyone involved – faculty and staff, current students, recent graduates, and conservation consultants – with exceptional exposure to the practical realities of emergency response, salvage, and long-term recovery efforts.

Lessons learned from Hurricane Katrina

Today, Heritage Emergency National Task Force members are working together on initiatives that apply important lessons learned from Hurricane Katrina. Incentives for preparedness, partnerships with first responders, training and coordination for regional events, disaster recovery funding and collaboration among museum and library organizations are top priorities. Task Force working groups are developing long-term recommendations in these areas, and they have also created several simple, new resources to help collecting institutions:

- Tips for working with emergency responders: getting to know local emergency responders and how they work before disaster strikes can help keep staff and collections safe. This tip sheet provides advice on how to find and build relationships with emergency responders, as well as what information needs to be exchanged to help responders protect cultural institutions.
- Guide to navigating FEMA and SBA funding: this concise website leads cultural institutions through the process of applying to the Small Business Administration (SBA) and FEMA for assistance after major disasters. Links put policies, application forms, and other necessary paperwork at the user's fingertips along with simple, step-by-step instructions for the application process.
- MayDay: MayDay is an annual event that encourages cultural institutions to undertake one simple emergency preparedness task. Created by the Society of American Archivists, MayDay was expanded in 2007 to include libraries, museums, and historic preservation organizations. In 2007, hundreds of organizations promoted the message, and FEMA featured MayDay on its website. The Task Force has produced a list of suggested MayDay activities, as well as promotional materials.

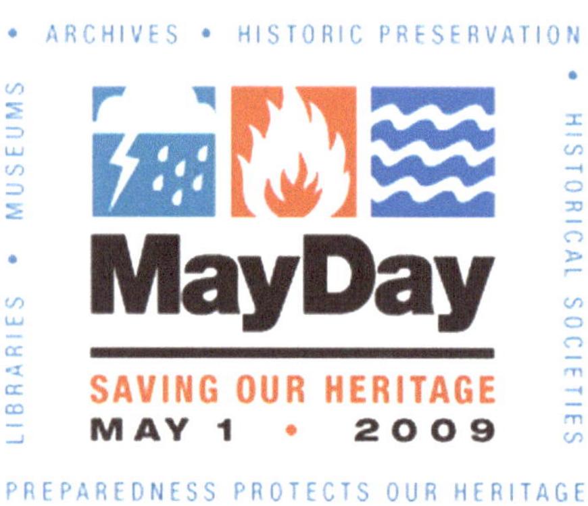

- Recommended professional emergency management training: most cultural stewards want to help in the aftermath of an emergency at their institution or in their community. The free courses listed in this resource teach cultural heritage staff and volunteer teams about local, state, and federal disaster response frameworks, as well as official protocols and structure, terminology, and key local contacts.

Despite significant progress many challenges remain. Working together, the Task Force[2] will continue to address urgent planning needs, foster stronger response networks, and conduct research on safe and efficient salvage protocols. We look forward to opportunities for international collaboration.

Notes

1. This paper is dedicated to my grandfather Primo Medici and to those who worked to preserve the treasures of Florence over 40 years ago and who inspired legions of future conservators, including one eighth-grade student (i.e. me) whose social studies report was fortunately focused on a July 1967 *National Geographic* article, 'Florence rises from the flood', on the recovery of art and artifacts in Florence.
2. Further information on the work of the Task Force and links to a variety of resources can be found at www.heritageemergency.org.

www.ingramcontent.com/pod-product-compliance
Ingram Content Group UK Ltd.
Pitfield, Milton Keynes, MK11 3LW, UK
UKHW061914230626
472545UK00003B/4
9781904982449